SpringerBriefs in Earth Sciences

More information about this series at http://www.springer.com/series/8897

T.N. Prakash · L. Sheela Nair
T.S. Shahul Hameed

Geomorphology and Physical Oceanography of the Lakshadweep Coral Islands in the Indian Ocean

 Springer

T.N. Prakash
L. Sheela Nair
T.S. Shahul Hameed
Coastal Processes Group
National Centre for Earth Science Studies
Trivandrum, Kerala
India

ISSN 2191-5369   ISSN 2191-5377   (electronic)
ISBN 978-3-319-12366-0   ISBN 978-3-319-12367-7   (eBook)
DOI 10.1007/978-3-319-12367-7

Library of Congress Control Number: 2014953291

Springer Cham Heidelberg New York Dordrecht London

Printed on acid-free paper

Springer is part of Springer Science+Business Media (www.springer.com)

Preface

Islands are often subjected to ecological, economic, and natural vulnerabilities. The Lakshadweep archipelago, a small group of coral islands in the Indian Ocean, is less prone to natural hazards such as cyclones, storm surges, tsunami, etc., due to its position. But the sea level rise, though a long-term threat, is an important natural hazard concerning the islands. Among the natural hazards, coastal erosion is the most serious problem faced by the Lakshadweep Islands due to their small size and topography. Erosion in the islands is caused by both natural and anthropogenic activities. The natural factors that contribute to erosion are high wave activity, strong winds, and currents, whereas the anthropogenic activities are mostly due to human intervention in the form of destruction of corals, construction of jetties, and other hard structures adopted including coastal protection. The erosion in the islands is also attributed to shifting beach sediments by the along-shore currents, energy concentration at certain segments due to wave diffraction, and reduction in the height of reef edge over a period of time. The relatively low land elevation of the islands makes them more susceptible to damage from high waves and flooding during adverse weather conditions. Though shore protection structures have been built in the islands by the Union Territory Administration, a long-term monitoring study is important to precisely identify the shoreline locations prone to erosion. A comprehensive study of the wave climate and coastal processes at work to delineate the factors responsible for shoreline changes and to identify the locations that need protection is needed. Simulation of the coastal processes can be effectively illustrated through numerical modeling. The impact of erosion on the islands can very well be demonstrated using this tool. Modeling results can provide vital information for the efficient formulation of disaster mitigation and management measures.

Energy requirement in the islands is met mostly through diesel. The fuel has to be transported from the mainland in large quantities and is stored in barrels, and in case of a spillage the sensitive island environment may be affected. Owing to fuel transportation, the cost of power generation is very high compared to the mainland. Non-conventional energy sources like solar, wave, and wind power can be alternative energy resources in the islands. Due to the geographical position of

the islands, solar energy is available throughout the year except during the monsoon period. During this period, wave and wind energy is at its highest which
could be tapped. In addition, the wave power potential of the Lakshadweep Sea is
higher compared to the coastal seas. Similarly, due to their exposure to the sea the
wind speeds on the islands are higher than on the mainland. Preliminary studies
on the economics of power indicate that the cost of wave/wind power generation
becomes comparable with the existing rates for which the fuel has to be transported from the mainland. A multisource power generation system is considered as
a technically and economically viable alternative source of energy for the islands.

Although the islands have long conjured up images of *'paradise'*, their amazing
lagoons and coral reefs show signs of increasing stress. The island communities
are striving to raise their living standards and as the population increases, there
is always a tendency to disturb the fragile ecosystem, which is one of the most
valuable assets as far as the islands are concerned. At times there is a tendency to
overexploit these natural resources and damage the environment. Another aspect
is the rising sea level due to global warming which is likely to damage the coastal
areas and even submerge some of the low-lying islands. This will certainly affect
the island economies with a negative impact on property, fisheries, tourism, coral
reefs, and freshwater resources. Islands are also important contributors to global
biodiversity as the lagoons and coral reefs are home to many rare species. There
are indications that these environmentally sensitive habitats are under increased
stress, which badly affects the flora and fauna of the islands, and in the case of
some of the native endangered species it may even lead to irreparable loss. To inte-
grate all these activities, an Integrated Coastal Zone Management (ICZM) plan is
required that would help to address the sustainable management of the islands.

T.N. Prakash
L. Sheela Nair
T.S. Shahul Hameed

सत्यमेव जयते

डॉ. शैलेश नायक
DR. SHAILESH NAYAK

सचिव
भारत सरकार
पृथ्वी विज्ञान मंत्रालय
पृथ्वी भवन, लोदी रोड़, नई दिल्ली-110003
SECRETARY
GOVERNMENT OF INDIA
MINISTRY OF EARTH SCIENCES
PRITHVI BHAVAN, LODHI ROAD, NEW DELHI-110003

FOREWORD

The Lakshadweep Archipelago, a group of numerous coral islands situated about 400 km off the west coast of India in the Arabian Sea has become strategically and economically, one of the most important regions in the country. These islands are unique for its aquatic bio-diversity with coral sands fringed by blue lagoon shallow waters having an Exclusive Economic Zone (EEZ) covering nearly 4 lakh sq.km. During the past couple of decades, there has been a sharp increase in the coastal developmental activities largely due to port and harbor development, shore protection and mainly recreation activities. These activities and their interactions with coastal processes have caused erotion at many places. This erosion is mainly attributed to high wave activity during the SW monsoon period.

The National Centre for Earth Science Studies (NCESS) (formerly CESS) has been conducting a number of studies for systematic collection of baseline data on erosion/accretion and wave measurements. This monograph resulting from the valuable studies is a significant contribution of beach changes, wave climate and coastal processes which will ultimately address the issue of shoreline changes and locations that need protection. It also projects the simulation of coastal processes which can be effectively done through numerical modeling. For the first time, an Integrated Coastal Zone Management (ICZM) plan was prepared for efficient formulation of disaster mitigation measures and sustainable management of the island coastline. In addition, the power situation in the islands is reviewed and a multi-source power generation system has been suggested as a technically and economically viable alternative source of energy in the islands.

I am sure this monograph will form a valuable baseline data on the Lakshadweep Islands for the planners, researchers and students. This is definitely a step forward for the overall development plan of the islands.

(Shailesh Nayak)

Tel. : +91-11-24629771, 24629772 ❏ Fax : +91-11-24629777 ❏ E-mail : secretary@moes.gov.in

Acknowledgments

We are grateful to Dr. Shailesh Nayak, Secretary, Ministry of Earth Sciences (MoES), Government of India for granting permission to publish and for writing the foreword to this book. We are highly indebted to Dr. N.P. Kurian, Director, National Centre for Earth Science Studies (NCESS), Trivandrum for providing the facilities and for critically evaluating the manuscript. We sincerely thank Dr. M. Somasundar from MoES who was officiating as Director-in-Charge for encouraging in taking up this publication. The authors are thankful to Dr. M. Baba, Former Director (Rtd.) for initiating many studies on Lakshadweep Islands. We would like to express our sincere thanks to the Administrator, UT of Lakshadweep, Government of India for sponsoring various studies on the Lakshadweep during different phases since 1989.

The following departments/organizations have helped with their commitment to the island people and with their tireless support during many years of research: Department of Science and Technology, Kavaratti, UT Lakshadweep; Ministry of Environment and Forests, Government of India; Indian Meteorological Department, New Delhi; National Institute of Ocean Technology, Chennai; ICMAM PD, MoES, Chennai; Chairman and Members, Dweep Council, UT Lakshadweep; Departments of Harbour Engineering, Electricity and Public Works, UT Lakshadweep; Technical Officers and Environmental Warden, DST, Kavaratti; M/s. Kapston, Planners and Builders, Aluva, Kochi.

There were many people recognized for their individual efforts for helping in many ways in this endeavor. Of special significance is the vision of Late Dr. K.R. Gupta, Advisor (Rtd.), Department of Science and Technology, New Delhi for giving much interaction and for inspiring us to take up this publication work on the Lakshadweep Islands.

This publication also recognizes many other individuals who made this work possible: Prof. Tad Murty, Adjunct Professor, University Ottawa; Dr. K.V. Thomas, Group Head, Coastal Processes Group, NCESS, Trivandrum; Dr. M.S. Syed Ismail Koya, Director (Rtd.), Department of Science and Technology, UT of Lakshadweep; Dr. T.K. Mallik, Director (Rtd.), Geological Survey of India; Dr. M. Prithvi Raj, Executive Secretary, Karnataka State Committee on Science and Technology, Bangalore; Late Dr. K.K. Ramachandran, Advisor (QA), CESS; Dr. M. Wafar,

Scientist (Rtd.), National Institute of Oceanography, Goa; Dr. A. Senthilvel, Director, MoEF, Government of India; Prof. Madhusoodana Kurup, Vice-Chancellor, Kerala University of Fisheries and Ocean Studies (KUFOS), Kochi.

In-house and field support provided by many staff members, especially Technical Officers Mr. D. Raju, Mr. Ajith Kumar, Mr. A. Vijayakumaran Nair, and Mr. M.K. Sreeraj and Research Scholars of NCESS, Mr. Tiju I. Varghese, Mr. V.R. Shamji, Mr. R. Prasad, Mr. Anish. S. Anand, Mr. R. Raveesh, Mr. S. Abhilash, Ms. Shinija Joseph, Mrs. Mrinal Sen and Mr. E.K. Sarath Raj have made this publication a success.

Contents

Chapter 1
Lakshadweep Islands

Abstract Lakshadweep Archipelago is a group of coral islands in the Indian Ocean. It has many islands and islets including submerged banks with a geographical area of 32 km². The islands are grouped into three clusters named as Laccadive, Aminidivi and Minicoy comprising of both inhabited and uninhabited islands. Here in this chapter we are presenting the general information about the origin, geology and geomorphologic setting, historical, socio-economic and ecological profile, resources of the islands, administrative set-up including the history of natural hazards.

Keywords Lakshadweep archipelago · Coral islands · Origin · Geomorphology and history of islands · Coral reefs · Natural hazards

1.1 Introduction

Lakshadweep, a group of coral islands located in the Arabian Sea off the west coast of India, forms an integral part of Chagos-Maldive-Laccadive Ridge in the Indian Ocean (Fig. 1.1). The Lakshadweep group of islands is the smallest Union Territory of India. It has 36 islands and islets consisting of 12 coral atolls, 3 reefs and 5 submerged banks lying between Latitudes of 8 and 12° 30′ North and Longitudes of 71 and 74° East [1] and is considered as a geographical extension of the Maldives Island chain further south. The geographical area of the entire group of islands put together is 32 km² with a coastline of length 132 km and lagoon area of 4,200 km². The islands are grouped into 3 main clusters named as Laccadive, Aminidivi and Minicoy and the islands comprising each group which include both inhabited and uninhabited are listed in Table 1.1.

Of the 36 islands, only 10 are inhabited viz., Agatti, Amini, Androth, Bitra, Chetlat, Kadamat, Kalpeni, Kavaratti, Kiltan and Minicoy (Table 1.2). All the islands in this group have a northeast-southwest orientation except the Androth which is characterized by an east-west orientation with no lagoon. All these islands have crescent-shaped banks with steep shores on the east and shallow

© The Author(s) 2015   1
T.N. Prakash et al., *Geomorphology and Physical Oceanography of the Lakshadweep Coral Islands in the Indian Ocean*, SpringerBriefs in Earth Sciences, DOI 10.1007/978-3-319-12367-7_1

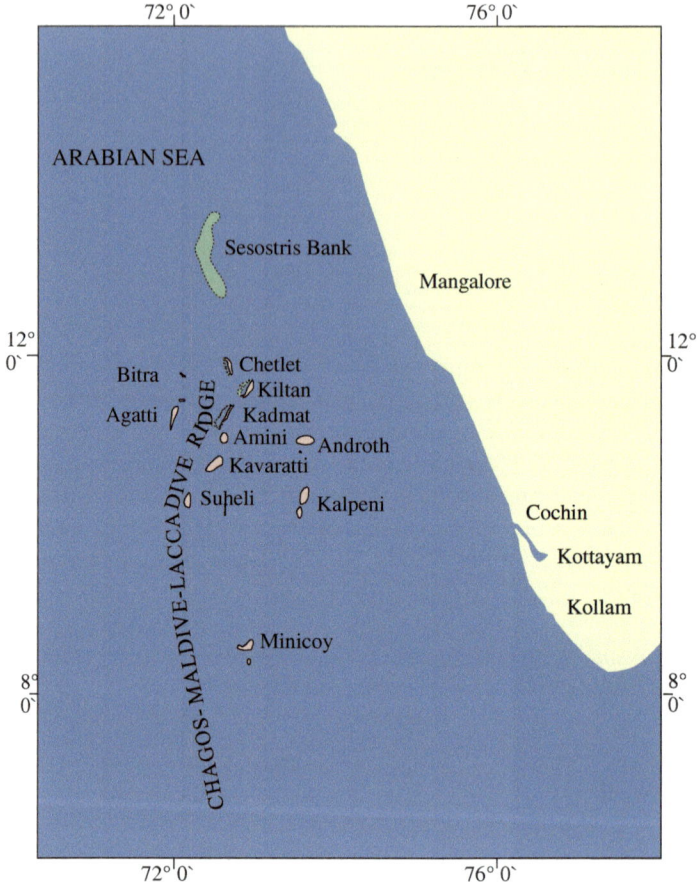

Fig. 1.1 Location map of the Lakshadweep group of islands

Table 1.1 Details of clusters of the islands

Cluster No.	Group name	Name of islands comprising the group
1.	Amindivi	*Inhabited* Amini, Kadmat, Kiltan, Chetlat and Bitra
2.	Laccadive	*Inhabited* Andrott, Kavaratti, Agatti and Kalpeni *Uninhabited* Kalpetti, Bangaram, Tinnakkara, Parali, Tilakkam, Pitti, Cheriyam, Suheli, Valiyakara, Pakshi Pitti and Kodithala
3.	Minicoy	*Inhabited* Minicoy *Uninhabited* Viringili

lagoons on the west. The reefs of these islands, come under atoll and platform types. The other associated features are reef flats, coralline shelf, coral heads, reef patches and live coral platforms, sand over reef and sand cays. The combined area occupied by the reef flat of all the islands is 136.5 km^2 [2]. Another typical feature

Table 1.2 Location of islands and its area

No.	Island	Land area (km^2)	Lagoon area (km^2)	Distance from mainland (Cochin) (km)	Population as per 2011 Census (No.)
1.	Agatti	3.84	17.50	459	7,560
2.	Amini	2.59	–^{a}	407	7,656
3.	Andrott	4.90	–^{b}	293	11,191
4.	Bitra	0.10	45.61	483	271
5.	Chetlat	1.14	1.60	432	2,345
6.	Kadamat	3.20	37.50	407	5,389
7.	Kalpeni	2.79	25.60	287	4,418
8.	Kavaratti	3.63	4.96	404	11,210
9.	Kiltan	2.20	1.76	394	3,945
10.	Minicoy	4.80	30.60	398	10,444

^{a}Lagoon sunken during geological past
^{b}No lagoon

Fig. 1.2 View of typical lagoon on the western side of the island

that is common to all these islands is the presence of a lagoon which is essentially a part of the submerged reef on the western side. The lagoons are of varying sizes and shapes, and in different stages of development. The lagoons are generally saucer shaped, wider at the middle and narrower towards the northern and southern ends (Fig. 1.2). Among the 10 islands Chetlat, Kiltan, Amini and Kadamat have smaller lagoons having a depth varying between 1.0 and 2.5 m, virtually filled with sediments. The bottom sediments of the lagoon comprise mainly of coral debris and calcareous sand [3]. Bitra, Bangaram, Suheli Par and Minicoy Islands have larger and deeper lagoons of depth up to 10 m. The area occupied by these lagoons varies from 1.6 to 46.25 km^2.

1.2 Origin of Islands

The evolution of the Lakshadweep Islands can be explained with the help of the theory put forward by Sir Charles Darwin, the renowned evolutionist. The origin of these islands can be traced to the gradual submergence of some of the volcanic ridge into the Indian Ocean followed by accumulation and growth of coralline deposits on the peaks and craters of these mountains (Fig. 1.3). These deposits then grew into coral islands resting on submerged mountain tops over a period of time. Each island is fringed by corals and is marked by a calm lagoon on the western side. The island along with the lagoon and its fringing coral reef form an atoll. The formation of atoll can also be explained in a simpler way. According to Darwin's theory, an atoll formation initiates with the development of a fringing reef around an island and gradually in due course the island gets submerged leaving only the ring shaped reef enclosing the lagoon. As per Murray's theory the atoll formation starts with the reef building process on top of a hill or plateau which has emerged from the ocean floor, wherein live coral on the outer edge has grown rapidly to reach the sea level first thereby forming a lagoon.

Fig. 1.3 Formation of an atoll on the volcanic plateau

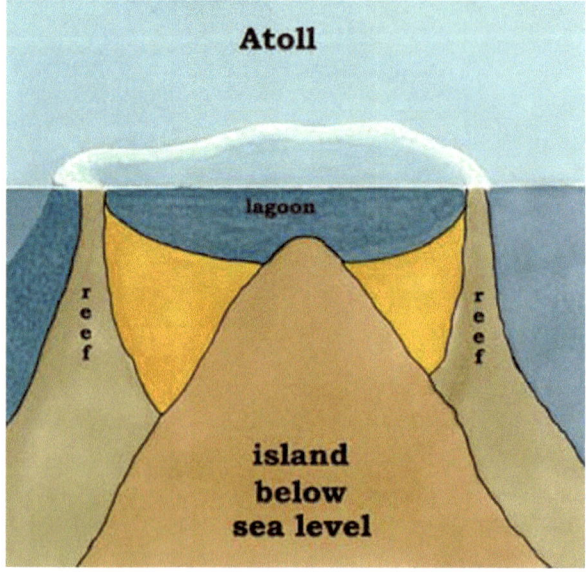

1.3 Geology and Geomorphology of Islands

The Chagos-Maldive-Laccadive ridge (Fig. 1.4) is an important linear submarine feature of the Arabian Sea [4] which extends in the north-south direction, for over 2,350 km, between 9°S at the southern end of Chagos Archipelago and 14°N around the Adas Bank. The ridges in the Lakshadweep Sea rise from a depth of 2,000 to 2,700 m along the eastern side and 4,000 m along the western side which consist of continental/transitional crust [5]. The eastern flanks of the ridge appear to be steeper compared to the western portion. It also has a number of gaps the prominent being the 9° channel which separates Minicoy (southernmost island of the group) from other islands of the Lakshadweep group.

The Lakshadweep archipelago is distinct with corals islands, banks and shoals, topographic rises and mounts, inter mountain valleys and sea knolls. The seaward reefs of the atolls have a number of well defined submerged terraces which have been attributed to the stand of sea level change during the Quaternary period [6]. Based on the structural features, trends of the individual islands, geophysical anomalies and related faults/dislocations, the Lakshadweep Islands are grouped into three blocks viz., northern, central and southern [7]. All the

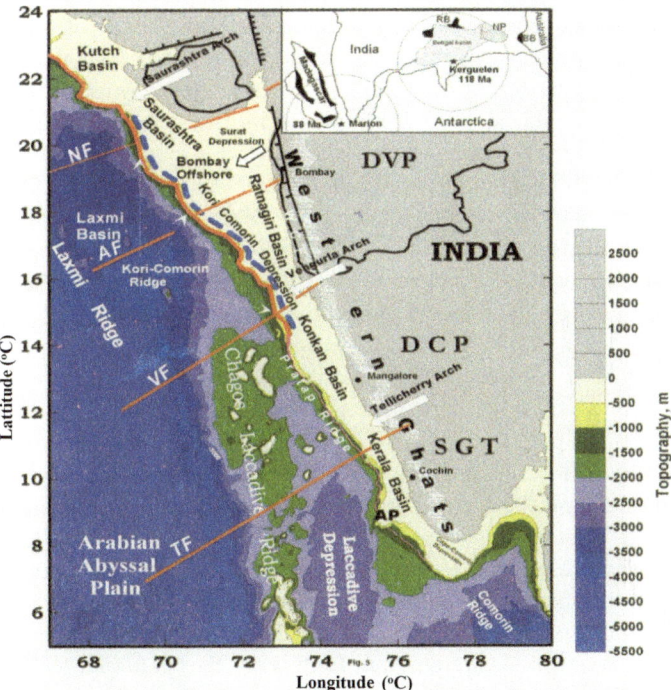

Fig. 1.4 Submarine geomorphology of the Arabian Sea (*Source* State of Environment Report for the Lakshadweep)

important islands fall in the central block separated by the Bassas de Pedro—a submerged bank towards north and a NNE-SSW trending valley in the south. The northern block comprises mostly of coral banks whereas the southern block has few islands and small banks. Topographic rises of the order of 500–1,300 m are observed in the slope region. At many sections, the slope has a faulted contact with abyssal plain [8].

All the islands have a surface layer of approximately 1–2 m thickness of coral debris and below that a compact but porous crust of limestone conglomerate. Below this, there is a bed of fine sand which acts as permeable layer through which there is constant filtration of fresh water. In general the sedimentary formation of Pleistocene to upper Paleocene age is encountered up to a depth of 300 m and below this depth volcanic rocks are present [9].

1.4 Historical Profile

Details on the early history of the Lakshadweep Islands are scanty. However, reference to the Lakshadweep can be seen in the memoirs of various travelers such as Marco Polo and Ibn Battuta who visited India during the 12th and 13th century respectively. These islands were considered as an important landmark by Vasco Da Gama, the Portughese explorer who first set foot on the west coast of India in 1497. Historical evidences show that towards the later part of the 17th century, the administration of the Lakshadweep was under Arakkal Bibi, the ruler of Cannanore in Kerala. In 1787 Bibi entered into a treaty with Tippu Sultan of Mysore and accordingly, the northern islands were transferred to the latter. But subsequently in 1799 Tippu Sultan was defeated by the British, and the Amindivi group of islands (northern group) came under direct control of the British [1]. In 1854–1855 the British attached the Laccadive Islands and Minicoy Island which was under the Ruler of Cannanore for not paying the taxes. In 1861 the Ruler of Cannanore cleared his dues to the British Government thereby releasing the attachment [1].

The early history of settlement in the islands is not recorded well. There are various legends which when compiled gives an overall idea about the historical processes and time periods. The first settlement on these islands was during the period of Cheraman Perumal, the last King of Chera dynasty, ruling the western Tamil Nadu and central Kerala. Cheraman Perumal later converted to Islam and left his capital Cranganore (the present day Kodungallur in Kerala) for Mecca without informing anybody. Parties were sent in different directions in sailing boats in search of him. One of the boats of the Raja of Cannanore was caught in a storm and wrecked on the Bangaram Island. Later, this group moved to the Agatti Island, located south of Bangaram, and from there they returned to the mainland. Another search party consisting of sailors and soldiers discovered the Amini group of islands and small settlements gradually started in Kavaratti, Androth and Kalpeni and later in the islands of Kadamat, Kiltan, Chetlat and Agatti [10].

1.5 Socio-economic Profile

The local population of Lakshadweep practices Islam religion. The total population of the Lakshadweep Islands as per census data of 2011 is 64,429 [11] with male and female numbering 33,106 and 31,323 respectively. Literacy rate is 93.15 % for males and 81.56 % for females. The language spoken is mostly Malayalam and a specialized dialect known as Jessri is also prevalent. In Minicoy, the people speak Mahl, which follows the Maldivian Dwehi script. The entire population of the Lakshadweep comes under the Scheduled Tribe (ST) category listed in the Constitution of India. The matrilineal system is being followed throughout the island where the land and property are inherited by the female in the family, which gives them a special status. The children are brought up in the mother's house. The concept of property sharing is also unique in the islands. Joint families are common where the families live together in their ancestral property known as '*tharavad*'. The tharavad is governed by the eldest male member (Karanavar) and can't be sold out. It can be mortgaged in lieu of a loan. The tharavad property varies from island to island. But gradually the joint family system is breaking up due to influx of modern nuclear family concepts leading to higher per capita land demand for settlement [12].

1.6 Livelihood

The main occupation of the islanders is fishing and coconut cultivation. Tuna is the major resource of fish followed by shark. The traditional pole and line method is generally adopted for tuna fishing (Fig. 1.5). The estimated annual potential of tuna resource is in the range 50,000–1,00,000 tons [13] while the same with respect to other fishes is estimated to be 25,000–50,000 tons. However, the highest exploited quantity till date is only around 12,000 tons with the tuna fish

Fig. 1.5 Catching tuna with pole and line from the Lakshadweep Sea

Fig. 1.6 Tourist huts and
other facility in the islands

accounting for 80 %. In addition, the Lakshadweep Sea also has potential for ornamental fishes, crustacean and molluscan resources, seaweeds and mariculture [14].

In addition to fishing, coconut cultivation is the other major occupation of the
people. The coconut yields oil, mira (sweet toddy) and jaggery all of which add to
the local economy. Small scale cultivation of other crops suitable for the islands is
also encouraged by the agricultural department by supplying vegetable seedlings
and other saplings through nursery. There are a few small scale industrial units
established for coir processing and allied works which use the coconut husk, that
are abundantly available in all the islands. The administration also imparts training
to islanders in making handicraft items like seashell-toys, coconut shell crafts and
wood carving using the locally available materials. These handicrafts are being
sold through khadi and village industrial units operating in the islands.

Tourism is another area where Lakshadweep has great potential. Agatti is the
only island where there is an Airport and is well connected to all the other inhabited
islands through speed vessels and passengers ships. Inter-island helicopter services
are also available. Government tourist huts and a few private hotels are available in
the major islands (Fig. 1.6). The Society for Promotion of Recreational Tourism and
Sports (SPORTS) is responsible for promoting the recreational activities linked to
tourism in the islands. A full-fledged Water Sports Institute exist at Kadamat Island
which attracts lot of visitors. The island administration is also considering the option
of opening up a few uninhabited islands for the promotion of tourism.

1.7 Administrative Setup

The Lakshadweep group of islands, which was earlier part of the erstwhile Madras
state of the Union of India, came into existence as the smallest Union Territory
of India on 1st November 1956 with Calicut in Kerala as its headquarters. The

headquarters of the Administration was shifted from Calicut to Kavaratti Island in 1964. The Administrator is the head of the Union Territory. The entire Union Territory of Lakshadweep is considered as one district with 10 sub-divisions of which 8 are controlled by Sub-divisional Officers and two (Minicoy and Agatti) by Deputy Collectors. The Lakshadweep Development Corporation set up under the Island Development Authority oversees the economic and commercial activities of the islands [15].

1.7.1 Panchayati Raj System

The Lakshadweep Island Councils Regulation 1988 and the Lakshadweep (Administration) Regulation 1988 are the acts under which the Island Councils and Pradesh Councils have been set up. The new Panchayati regulation implies the formation of a two tier system of Panchayats in the Lakshadweep. There are Dweep Panchayats and a District Panchayat. The ten inhabited islands have 10 Dweep Panchayats. The district Panchayat has a Chairperson and a Vice Chairperson. The Chairpersons of the Dweep Panchayat are also members of the District Panchayat. The Panchayats discharge their functions as provided in the Lakshadweep Panchayat Regulation 1994. The Village (Dweep) Panchayats were constituted in December 1997 followed by the District Panchayat in January 1998. Subsequent to this the Lakshadweep Administration transferred schemes and programmes from various important sectors like Agriculture, Animal Husbandry, Public Health, Industries, Cooperative Societies, Education, Fisheries, Social Welfare, Electricity, Environment, Rural Development, Public Works etc. along with the necessary finance and manpower to empower the Panchayats (Website: http://www.lakshadweep.nic.in).

1.7.2 Judicial System

The high court of Kerala exercises judicial supervision over the courts in the Union Territory of Lakshadweep. The District Magistrate who has jurisdiction over all the islands is assisted by one Additional District Magistrate and ten Executive Magistrates with respect to the enforcement of law and order. The Lakshadweep police force is under the control of the Administrator who is also the Inspector General of Police.

1.8 Topography and Surface Characteristics

Most of the islands of the Lakshadweep archipelago are built on a submerged volcanic platform. The islands are narrow, encircled by a beach consisting of calcareous sands. Generally the elevation of the islands ranges from 0.5 to 6.0 m above mean sea level. The interior parts of the islands have calcareous sand

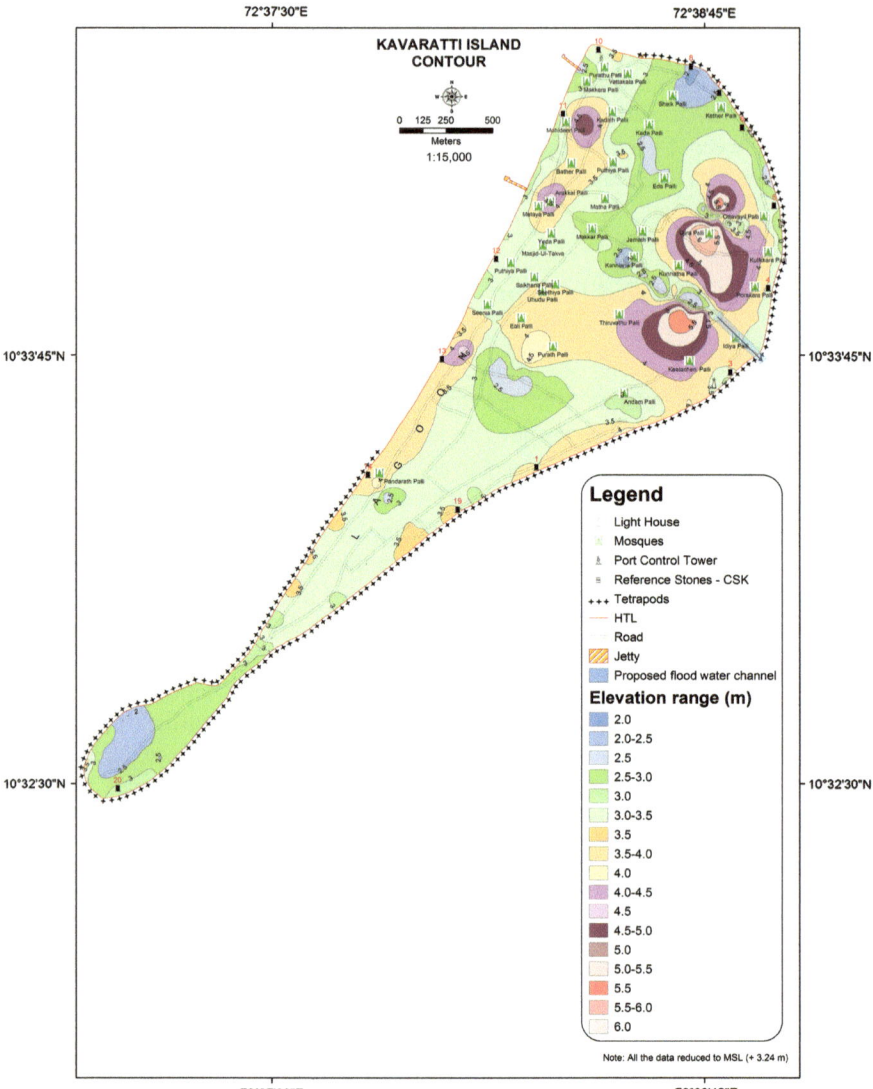

Fig. 1.7 Topography of the Kavaratti Island

deposit, which could have been altered by anthropogenic activities. The elevation of the sand dump/sand dune is as high as 6 m above MSL and is separated by saddle portions, which almost touch the ground water table supported by a horizontal basal layer of coral limestone at a depth of 0.5–1.5 m from the surface [16]. Figure 1.7 gives the topography of one of the islands.

The top soil in the islands hardly contains any organic matter. As the top soil is formed by disintegration of coral limestone underlined with pebbles of different shapes and sizes, it is deficient in micro and macronutrients essential for plant growth. The pH of the soil is almost neutral, ranging from 6 to 8. It is estimated to contain 95 % calcium carbonate in the form of aragonite. Organic carbon content ranges between 0.8 and 2.2 %. These soils have poor water holding capacity and are extensively used for coconut plantations. The color of the soil varies from light yellowish brown to light brown and from light grayish brown to grey.

1.9 Groundwater Resources

Fresh water resources in the islands are limited. The hydrological regime is extremely fragile. The depth to water table varies generally between 0.5 and 3.5 m below ground level. The water table slope is steeper towards the lagoon side than towards the open sea on the eastern side. Tidal influence is seen in the well of all the islands. The water is periodically renewed by rainfall. It is observed that the over exploitation of water resources are continuing, resulting in drastic lowering of ground water table which invariably leads to saline intrusion [17]. The island population used to depend on bore wells and surface wells to meet their water demand. Rainwater harvesting is also being adopted in all islands in a big way as it is the purest source of potable water available in all the islands and copious quantities are available during the monsoon period which can be stored. But the quantity available during the monsoon period is not sufficient to meet the demand during the non-monsoon months. Hence the administration is exploring other options like desalination. The first desalination plant of 1 lakh litre/day capacity was installed at the Kavaratti in 2005 followed by similar plants at the Agatti and Minicoy in 2009. All the plants are indigenously developed by the National Institute of Ocean Technology, Ministry of Earth Sciences, Government of India based on Low Temperature Thermal Desalination (LTTD) technology.

1.10 Ecological Profile of the Islands

The marine biodiversity of the islands is predominantly related to the diversity of the coral reefs (Fig. 1.8). The coral diversity of the Lakshadweep is second to that of the Andaman and Nicobar Islands (Bay of Bengal Sea off east coast of India). Coral reefs and the islands that are formed due to the accretionary processes are vital both in the economic and ecological sense. The ecosystems on the islands and the surrounding seas can be broadly classified as: (i) open sea, (ii) coral reef, (iii) lagoon, (iv) beach (lagoon and open sea) and (v) Island.

Alcock [18] visited several islands of the Indian seas and recorded the different corals for the first time. The earliest detailed studies on the reefs of the Lakshadweep are by Gardener [19] wherein the corals and different reef systems are described. Pillai [20, 21] studied the Minicoy atoll for over a decade. A team of

Fig. 1.8 Coral reefs of the
Lakshadweep

scientists from the Central Marine Fisheries Research Institute (CMFRI), Cochin;
National Institute of Oceanography (NIO), Goa; and the Zoological Survey of
India (ZSI) have carried out several studies in this region. Survey of faunal assem-
blages of the Lakshadweep was undertaken by ZSI during 1982–1987 and in 1991.
Similarly CMFRI conducted studies on the fishery potential and the findings were
published in a special issue of CMFRI bulletin (1989) on the Lakshadweep. NIO
has published the atlas of corals and coral reefs of India in 1986.

1.10.1 Coral Reef

Lakshadweep has a total reef area of 16.1 km^2 covering 12 atolls, 3 reefs and 5
submerged banks. Most of the islands are located on the windward reef flat in the
eastern side. The windward reef flats have well developed algal ridges. The lee
side fringes are mostly devoid of any live corals. However the rock pools are seen
to house isolated coral colonies. The windward reef flats have huge limestone col-
lections which show signs of erosion due to high wave activity. Coral genera like
Montipora, Pavona, Porites, Favia, Favites, Goniastrea, Platygyra, Hydnophora
and Symphillia are common here [22]. Some sub-genera like Psammocora
(Plesioseris) and Psammocora (Stephanaria) are also present. On the lagoon
shoals and the windward and leeward sides of the reefs, genera like Pocillopora,
Acropora, Porites, Goniastrea among the scleractinians and the blue coral
Heliopora are found. The reefs and shoal of the Minicoy and Chetlat Islands have
hemispherical colonies of blue coral which occupy nearly 80 % of the reef surface
and lagoon floor. Psammocor is another common genus found particularly in the
Kadamat and Chetlat Islands. Species like Lobophyllia and Diploastria are present
in the lagoon of Minicoy which are also found in nearby the Maldives Islands. A
total of 78 species of Scleractinian corals divided among 31 genera are hitherto
reported from the Lakshadweep. Out of these 27 genera with a total of 69 species
are hermatypes and the rest 4 genera with 89 species are ahermatypes [22].

Table 1.3 Status of live corals in the islands

Sl. No.	Island	Status[a]
1.	Agatti	+++
2.	Amini	+
3.	Andrott	--
4.	Bangaram	++
5.	Bitra	+++
6.	Chetlat	++
7.	Kalpeni	--
8.	Kavaratti	++
9.	Kadamat	--
10.	Kiltan	+++
11.	Minicoy	--
12.	Suheli	++

[a]Unsatisfactory: -- (10–15 %); Satisfactory: + (above 15 %); Good: ++ (above 20 %); Very good: +++ (above 30 %)

1.10.2 Status of Coral Reefs

The Lakshadweep atolls, even though inhabited, remained protected due to the restricted access. However the changing demographic pattern and life style in the recent years coupled with resource harvest from the reefs have put enormous pressure on the coral ecosystem. Studies on the status of coral reef in the Lakshadweep based on ground level observations were conducted by Pillai [20, 21] and Pillai and Jasmine [23]. Also direct monitoring of the corals at selected islands was carried out by engaging divers belonging to different organizations. The current status of live Coral Reefs in the Lakshadweep are compiled and presented in Table 1.3.

1.11 Natural Hazards

Generally, the islands of the Lakshadweep are least affected by natural calamities like earthquakes, storms, cyclones due to its geographical position and general orography. According to the available information, earthquakes have not occurred in this region. The 2004 tsunami which wrecked havoc along the coasts of mainland and Indian Ocean rim countries did not have any impact on the islands. Only seven storms have affected the Lakshadweep Islands in the past 166 years (Table 1.4). Earliest record of natural calamity in the Lakshadweep area is that of the great storm that occurred in April, 1847. There is also a record of a violent storm which hit the Kavaratti Island in 1891 causing extensive damage to coconut trees. There are also reports of the same storm hitting Agatti and its attached islets, in the Laccadive group and the Amindivi group of islands. The most recent storm that hit the island was in May 2004 [24]. The islands of Amini Kiltan, Agatti and Kavaratti were affected (Fig. 1.9a, b). Whenever a cyclone or a storm hits the islands, the first causality is the coconut trees and the fruit

Table 1.4 Storms that affected the Lakshadweep Islands

Year	Loss of life	Severely affected islands	Other affected islands
1847	Kalpeni, 246 deaths	Kalpeni, Androth	Kiltan
1891	Property loss	Androth	Kavaratti, Agatti, Amini and Kalpeni
1922	No loss of life	Kalpeni	
1963	No loss of life	Androth, Kalpeni	Agatti, Kiltan
1965	No loss of life	Androth, Kalpeni	Androth, Kalpeni, Agatti, Kiltan
1977	No loss of life	Kalpeni	Androth
2004	Property loss	Amini	Kiltan, Kavaratti, Agatti

Source Plan Document, UTL

Fig. 1.9 Flooding and devastation during 2004 Cyclone **a** Amini Island; **b** Kavaratti Island (*Source* Prakash and Hameed 2004)

(a)

(b)

bearing trees followed by disruption of fishing activities affecting the livelihood of the local people. These events not only affect the physical and social infrastructure, but also slow down the pace of development in these islands.

Recent studies also indicate that the shoreline of many islands has been receding which is a matter of great concern to the administration. In majority of the islands the per capita land availability is less leaving very little space for further development and also for planning and implementation of disaster mitigation measures. The administrative authorities are taking all possible measures to conserve the existing land area by adopting suitable land use practices. A detailed scientific study of the shoreline morphology and its temporal and spatial variations along with numerical modeling to understand the coastal processes at work is needed for the development of an efficient coastal management system. The baseline studies carried out in all the three group of islands which includes beach morphology, hydrodynamics and modeling of coastal processes are presented in the subsequent chapters. Different aspects of coastal management issues including the potential for non-conventional renewable energy sources are also addressed.

References

1. Mannadiar NS (1977) Gazetteer of India (Lakshadweep). Government Press, Coimbatore
2. Anjali B, Shailesh N (1994) Coral reef mapping of the Lakshadweep Islands. Scientific Note: SAC/RSA/RSAG/DOD-COS/SN/09/94. Space Application Centre, Ahmadabad
3. Mallik TK (1976) Grain-size variation in the Kavaratti lagoon sediments, Lakshadweep, Arabian Sea. Mar Geol 20:57–75
4. Narain H, Kaila KL, Varma RK (1968) Continental margins of India. Can J Earth Sci 5:1051–1065
5. Naini BR, Talwani M (1982) Structural framework and the evolutionary history of continental margin of India. In: Watkins JS, Drake CL (eds) Studies in continental marine geology. Am Assoc Petrol Geol Mem 34:167–191
6. Siddiquie HN (1975) Submerged terraces in the Laccadive Island. Mar Geol 18:M95–M101
7. Zutshi PL, Murthy MSN, Thakur SS (2001) Exploration for Hydrocarbon in and around Lakshadweep Islands. Geol Surv Ind Spl Pub 56:59–69
8. Shrivastava JP, Nair KM, Ramachandran KK (1978) Reports on the geological expedition to Lakshadweep Islands. Oil and Natural Gas Commission, Geosciences Division, Calcutta (unpublished report)
9. Biswas SK (1987) Regional tectonic framework, structure and evolution of the western continental basins. Tectonophysics 35:307–327
10. Mukundan TK (1979) Lakshadweep—A hundred thousand islands. Academic Press, New Delhi
11. Anonymous (2011) Census of India, 2011. Office of the Registrar General and Census Commissioner, India
12. Ramunny M (1972) Laccadives, Minicoy and Amini Islands. Publications Division Govt of India, New Delhi
13. Integrated Coastal Zone Management Plan (ICZMP) (2006) Final report, Submitted to Ministry of Environment and Forests, Govt of India, Centre for Earth Science Studies, Trivandrum
14. Silas EG, Pillai PP (1982) Resources of tuna and related species and their fisheries in the Indian Ocean. CMFRI Bull. ICAR, Cochin, India
15. Union Territory of Lakshadweep (2007) Draft Eleventh Five Year Plan (2007–2012), Planning and Statistics Department, UT Lakshadweep, Kavaratti
16. Muralidharan MP, Praveen Kumar P (2001) Geology and geomorphology of Kavaratti Island Union Territory of Lakshadweep. Geol Surv Ind Spl Pub 56:9–13
17. Varma AR, Unnikrishnan KR, Ramachandran KK (1989) Geophysical and hydrological studies for the assessment of ground water resource potential in Union territory of Lakshadweep, India. Final report, Centre for Earth Science Studies, Trivandrum

18. Alcock A (1893) On some newly recorded corals from the Indian Seas. J Asiat Soc Bengal (Nat Hist) 62(II):138–149
19. Gardiner JS (1903) The Maldives and Laccadive groups with notes on other coral formations in the Indian Ocean. In: Gardiner JS (ed) Fauna and geography of the Maldives and Laccadive Archipelagoes. Cambridge University Press, Cambridge
20. Pillai CSG (1971) Distribution of shallow water stony corals at Minicoy Atoll in the Indian Ocean. Atoll Res Bull Washington 141:1–12
21. Pillai CSG (1971) Composition of the coral fauna of the southern coast of India and the Laccadives. In: Yonge C, Stoddart DR (eds) Regional variation in Indian Ocean coral reefs. Symposium of the Zoological Society of London. Academic Press, London
22. DST (2002) Biophysical surveys (1999–2002). UT Lakshadweep
23. Pillai CSG, Jasmine S (1989) The coral fauna marine living resources of the UT of Lakshadweep. CMFRI Bull 43:179–194
24. Prakash TN, Shahul Hameed TS (2004) Site inspection report (Cyclone-2004) in Lakshadweep Islands. Submitted to UT Lakshadweep, Centre for Earth Science Studies, Trivandrum

Chapter 2
Hydrodynamics of Lakshadweep Sea

Abstract The winds in the Lakshadweep are predominantly northwest throughout the year with higher speeds in the range 5–8 m/s during June through August, the southwest monsoon season. The wave climate of the Lakshadweep Sea is dominated by the southwest monsoon. The highest wave heights are observed during the period from June to August with the dominant values of maximum wave heights around 5 m. During the rest of the period it was around 1.4 m. Higher wave periods are observed in February and April and the lower ones during June–August. The dominant zero-crossing period during the southwest monsoon is in the range 7–8 s and in the rest of the months it varies from 5 to 10 s. South-southwesterly waves dominate during most of the periods except during the peak southwest monsoon when west-southwest to westerly waves dominate. The inner-shelf currents are generally weak with northeast to southwest directions. The tide in the Lakshadweep Sea is mixed, semidiurnal and micro-tidal type.

Keywords Lakshadweep Sea · Waves · Currents and tides · Wind · Significant wave height · Wave direction · Wave climate

2.1 Introduction

Hydrodynamics generally refers to the study of the dynamics of fluids. It deals with the flow as well as the characteristics of the fluid in motion, in this case the sea water. Thus the characteristics of the wind, waves, currents and tide in the coastal waters of the Lakshadweep are explained here. They are interlinked in one way or other. For example, the waves are generated mainly due to the action of wind blowing over the sea surface. The dynamic energy is transferred to the sea surface nonlinearly by the wind. The wind is one among the driving force for the ocean currents also. On the other hand, the sea is transferring energy to the wind, as in the case of storms/cyclones moving above the sea. Finally, it is by the action of the waves and currents the sediment transport is taking place.

© The Author(s) 2015
T.N. Prakash et al., *Geomorphology and Physical Oceanography of the Lakshadweep Coral Islands in the Indian Ocean*, SpringerBriefs in Earth Sciences,
DOI 10.1007/978-3-319-12367-7_2

2.2 Waves

Centre for Earth Science Studies (CESS) under the financial assistance from the Department of Ocean Development, Government of India carried out wave measurements in the sea off Kavaratti [1] by deploying a directional wave rider buoy during March 1991–March 1992 [2, 3]. The buoy was deployed off Kavaratti on the southern side at a depth of 30 m. Due to the very steep bathymetry of the shelf the shoaling effects at this point is negligible when compared to the deep water waves. Hence, the waves recorded at this point can be taken as deep water waves for all practical purposes. In this section, the characteristics of the waves are presented.

2.2.1 Wave Heights

The parameters of wave height presented in this section are the maximum value of wave heights in each recording period, usually known as the maximum wave height (H_{max}) and the average of the highest one-third of the wave heights in each recording period, known as the significant wave height (H_s). The H_s is mostly used in engineering applications since it reflects the energy content of the wave group.

The maximum wave height (H_{max}) during each recording period ranged from 0.56 to 8.95 m, with the lowest values observed in February and the highest in August. Percentage occurrence of H_{max} during the different months are presented in Fig. 2.1 which shows that H_{max} is generally less than 5.0 m during November–March. During the above period the predominant H_{max} is in the range 1.2–1.6 m, which constitutes nearly half of the cases. From June onwards the wave intensity increases and higher waves persist for 3 months. During June–August (monsoon season) the wave heights reach its maximum as expected and the distribution is flatty kurtic. The highest occurrences are in the range 4.8–5.2 m which contributes 21 % and the range 4.0–5.2 m constitutes 51 % during June–August.

The monthly average H_{max} in each month is given in Fig. 2.2. The highest wave activity is observed during June–August, as seen in the monthly distributions. The average H_{max} during these months are in the range 4.1–5.0 m, the highest occurring in July. During the other months the average values range from 1.3 to 2.7 m. During the months of November–March, lower values of less than 2 m are observed.

The significant wave height ranged in the year from 0.4 to 4.7 m, the lowest being observed in February and the highest in August as in the case of H_{max}. The monthly distributions of H_s are depicted in Fig. 2.3 as percentage occurrences. The maximum values of H_s are observed during June–August with ranges 2.4–2.6, 3.0–3.2 and 2.0–2.2 m contributing about one-fifth each of the total occurrence. The wave activity decreases since then and the peak of the distribution of H_s during September and October are in the range 1.4–1.6 m. The wave activity is still less intense during November–March. Maximum occurrences of H_s during these months are in the range 0.8–1.0 m with contributions between 40 and 50 %. During March, November and December the wave activity is slightly more intense compared to January and February with H_s in the range 0.8–1.2 m contributing about 65–70 % of the occurrence. During April and May the

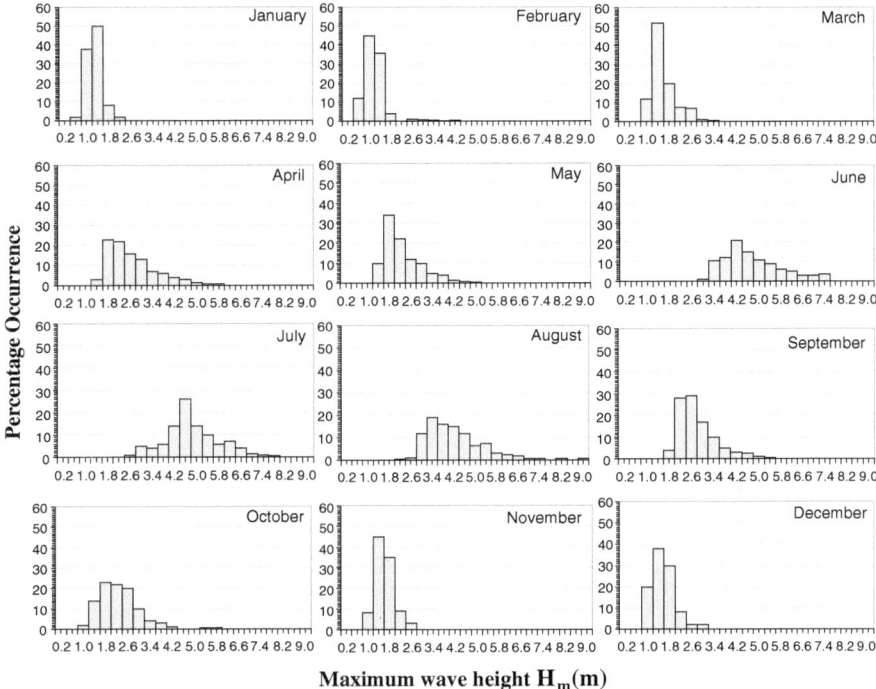

Fig. 2.1 Monthly distributions of maximum wave height (after Baba et al. [1])

Fig. 2.2 Monthly average of maximum wave height (after Baba et al. [1])

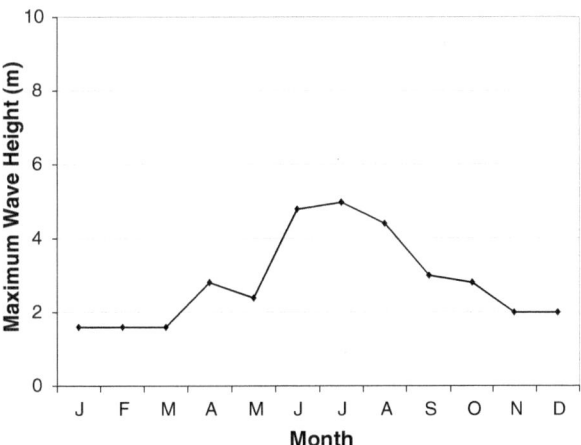

waves are slightly higher. The monthly mean H_s in each month is given in Fig. 2.4. High values of H_s are observed, as expected, during the months of June, July and August. During these months the monthly mean value of H_s varied from 2.5 to 3.0 m. During the other months mean H_s are in the range 0.8–1.7 m. The lowest wave heights are observed during November–March. During these months, the mean H_s ranged from 0.8 to 1.2 m.

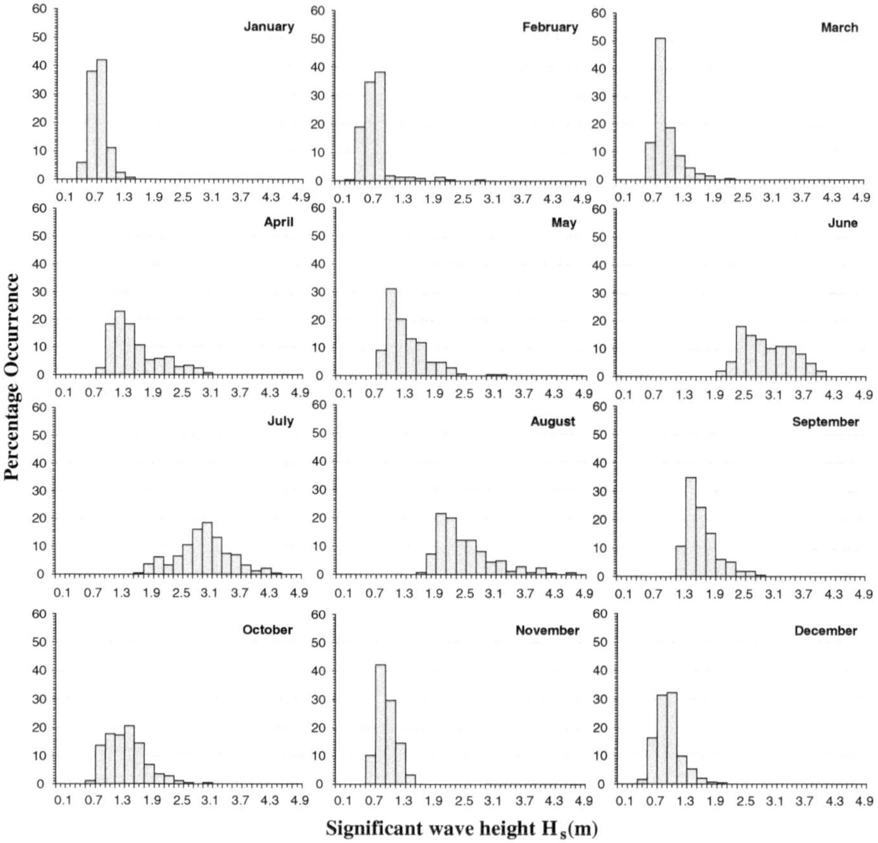

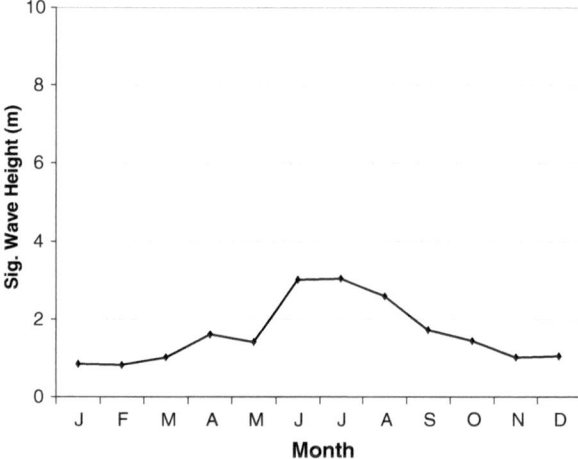

Fig. 2.3 Monthly distributions of significant wave height (after Baba et al. [1])

Fig. 2.4 Monthly average of significant wave height (after Baba et al. [1])

2.2.2 Wave Periods

The usually referred wave period parameters are the zero-crossing period and the peak period. The zero-crossing period (T_z) is the time taken by two successive crests to cross the zero or mean water level. It represents the average period of the waves in a wave train. Similarly, the period corresponding to the maximum wave energy spectrum is referred as peak period (T_p). It is mostly used in engineering applications at it represents the peak energy of the wave train.

The zero-crossing period vary from 3.5 to 13.3 s annually. The maximum range is observed in February and April and the minimum during the monsoon period of June–August. The monthly percentage occurrence of T_z (Fig. 2.5) shows that during January and February T_z in the range 5.0–6.5 s constitute more than half of the distributions. During March and April the waves are poorly sorted. During March there are two peaks in the distribution of T_z with nearly equal occurrences in the ranges

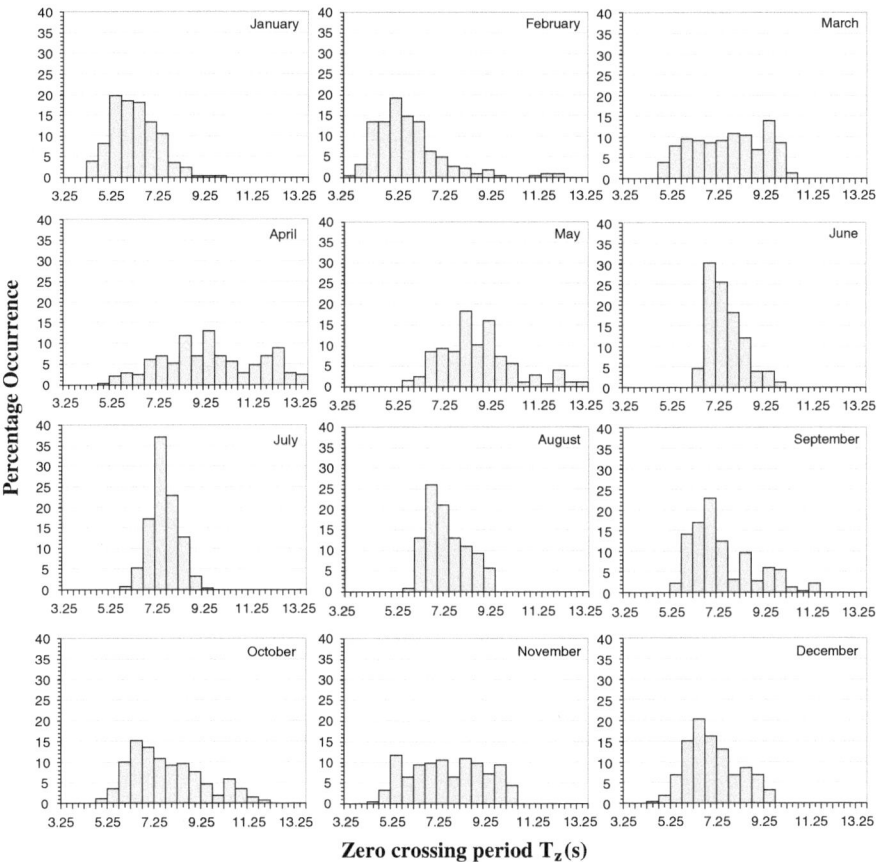

Fig. 2.5 Monthly distributions of zero crossing period

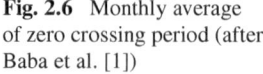

Fig. 2.6 Monthly average
of zero crossing period (after
Baba et al. [1])

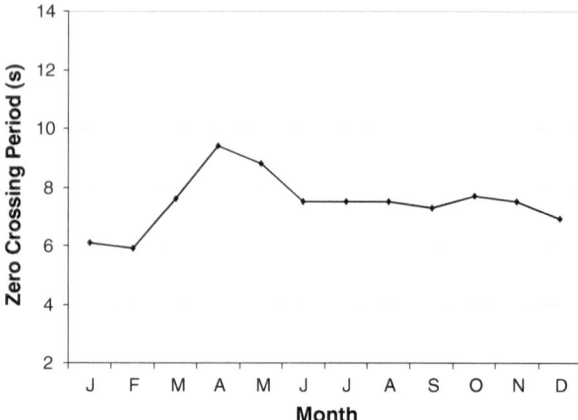

of 5.0–6.5 s and 8.0–9.5 s. During April the peak of the distribution is in the range
8.5–9.5 s with about one-fourth of the occurrence. During May the wave periods are
comparatively higher with about half in the range 8.0–9.5 s. During June–August
waves are comparatively better sorted with the peak distribution falling in the range
7.0–7.5 s constituting about one-fourth to one-third of the distributions. During June
T_z in the range 6.5–7.5 s constitutes more than half of the cases. During July and
August more than three-fourth of the distribution is confined to the range 7.0–8.5 s
and 6.0–8.0 s respectively. From September onwards the range of the distribution
is widened. During September T_z in the range 5.5–7.5 s constitutes about three-
fourth during this month. During October and November the distributions become
more flatty-kurtic. During October T_z in the range 6.0–7.0 s represents one-fourth
and that in the range 5.5–9.0 s represents three-fourth of the distribution. During
November one-fourth of the distribution is in the range 8.0–9.0 s whereas during
December 60 % of the distribution is in the range 5.5–7.5 s.

The monthly average values of T_z are depicted in Fig. 2.6. The monthly mean
T_z ranged from 5.9 to 9.4 s. During June–November the values are almost uniform
and are around 7.6 s. However, during September–November the range of occurrence are more compared to June–August. The lower range and better sorting during the monsoon season confirms the proximity of the generating zone.

The peak period (T_p) ranged from 8.4 to 26.0 s during the year, the lowest being
observed in November and the highest in February. Generally, T_p is lower during
June–August with small variations. The monthly distributions of T_p (Fig. 2.7) show
that the distributions are somewhat similar in pattern, to the distribution of T_z. The
maximum spread with lower percentage occurrence of T_p is observed in January
and February. During January T_p in the range 11–18 s constitutes nearly 80 % of
the distribution, whereas during February, T_p in the range 14–21 s represents three-
fourth of the distribution. During March the peak is in the range 14–15 s contributing to one-fourth of the distribution and T_p in the range 14–18 s represents
nearly three-fourth of the distribution. During April T_p values are slightly higher
with nearly half the cases in the range 15–17 s. During May, more than 80 % of

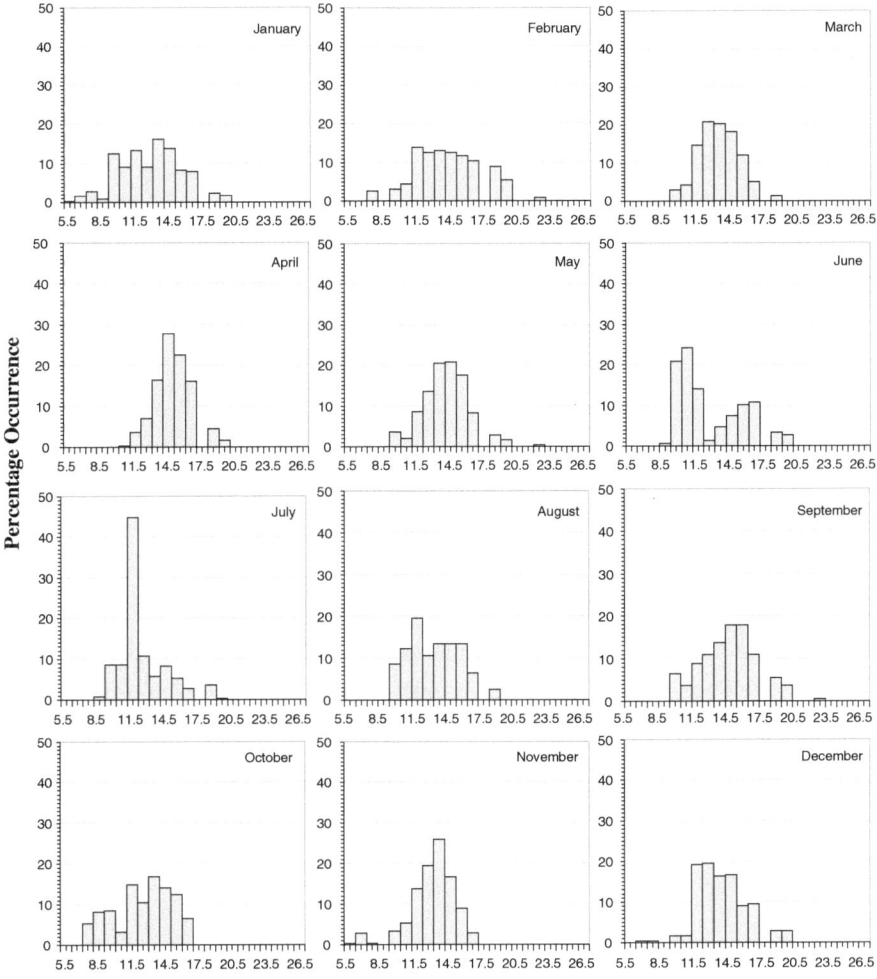

Fig. 2.7 Monthly distributions of peak period (after Baba et al. [1])

the T_p values fall in the range 14–18 s. By June, the values become lower with lower spread, with the onset of the southwest monsoon, when the waves are better sorted. During June, July and August the peaks of the distributions are in the ranges 11–12 s, 12–13 s and 14–15 s with about one-fourth of the contributions (24, 29 and 30 % respectively). During June T_p in the range 11–15 s represents more than two-third of the distribution. During July and August more than half of T_p fall in the range 12–14 s and 13–15 s respectively. The increase in T_p is continued in September with the shifting of the peak in the range 15–16 s. During September the distribution is widened and nearly half of the values fall in the range 14–17 s. During October and December the peaks of the distributions lie in the range 14–15 s with nearly one-fifth of the values falling in this range. During

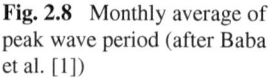

Fig. 2.8 Monthly average of peak wave period (after Baba et al. [1])

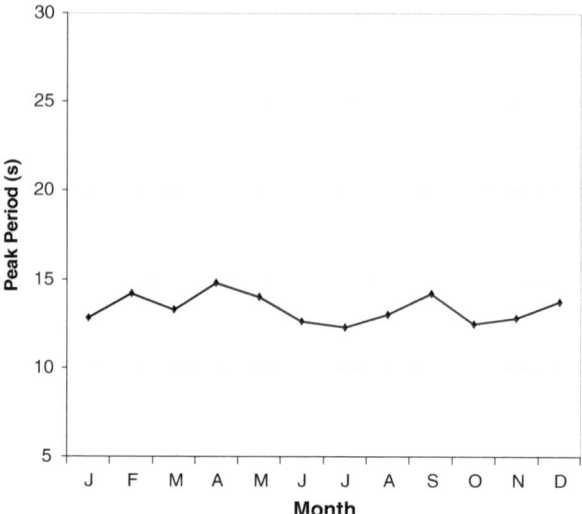

October the distribution is further widened with nearly three-fourth of the values falling in the range 13–17 s and during November the ranges 13–14 s and 14–15 s have equal frequencies, both together constituting about 40 % in the distribution. During December about two-third of the values are found in the range 13–17 s.

The monthly mean values of T_p are depicted in Fig. 2.8. The values range from 13.6 s in June to 17.9 s in February. The lowest values are observed during June, July and August, as expected. The characteristics during June–August, as observed from the distributions of T_z is confirmed from the distribution of T_p also. During these months the average values vary between 13.6 and 14.3 s only. During February–May and September the mean T_p is between 16 and 17 s.

2.3 Wave Direction

The wave direction corresponding to the peak spectral density during each recording period (D_p) is dealt here because it corresponds to the direction of the major part of the wave energy. It is seen that D_p ranges from 106 to 316°N during the year. The monthly distributions of wave directions (D_p) are presented in Fig. 2.9. During most of the months, except during the monsoon months, the peaks of the distributions are around 200°–210°. The dominance of the peak varies with months. While the peak at 200°–210° represents one-fourth the cases in January, it is nearly half in February, April and May, one-third in March and about two-fifth in September, November and December. During June–August the distribution give a different pattern. During June the peak of the distribution is in the range 250°–260° with 20 % contribution followed by a secondary peak with 15 % occurrence in the range 200°–210°. During this month directions in the ranges 230°–270° and 190°–220°

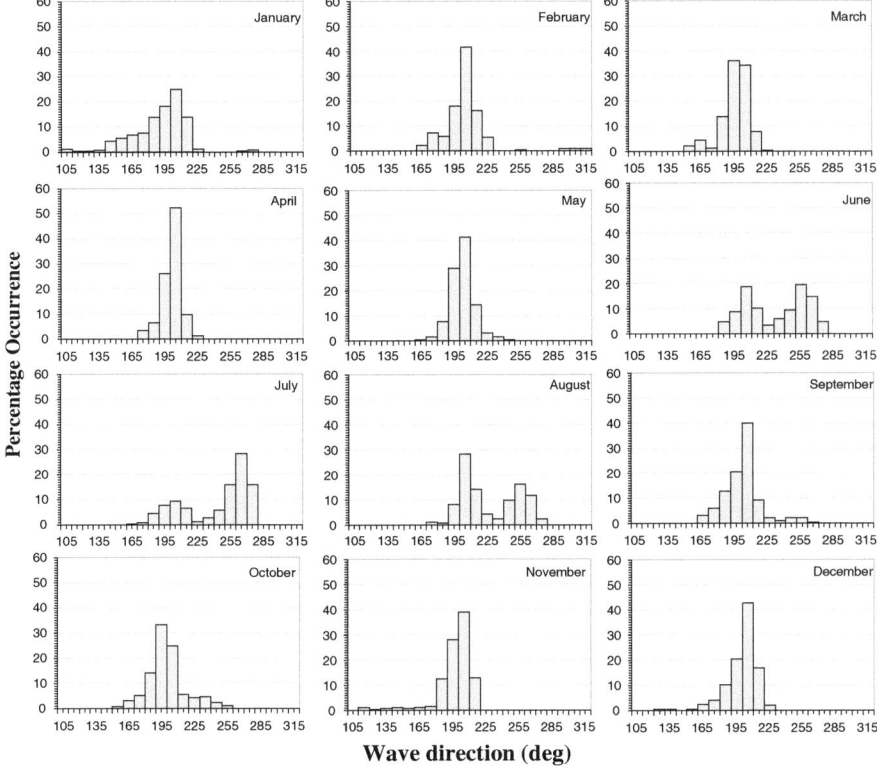

Fig. 2.9   Monthly distributions of wave direction (after Baba et al. [1])

represents more than half and one-third respectively in the distribution. During the month of July the dominant direction is near to west and the peak is at 260°–270° with 30 % contribution. During this month about two-third of the distribution is in the range 250°–280°. During August the distribution shows a pattern similar to that observed in June, but with more prominence to the direction in the range 200–210 with more than one-fourth of the cases. Directions in the ranges 200°–220° and 240°–270° represents nearly half and one-third of the distribution during this month. By September the westerly components are still reduced and the southerly directions become prominent, as noted in the beginning of this section. During October the peak lies in the range 190°–200° with nearly one-third of the contribution and those in the range 190°–210° represent 60 % of the distribution. Thus, it follows that the southerly-southwesterly waves are persistent in the Lakshadweep throughout the year and the westerly waves dominate only when the monsoon is very intensive. Similar observations are also reported [4] for the southwest coast of India.

The monthly mean wave directions are presented in Fig. 2.10. The mean wave directions range from 191 to 246°N. The waves are least persistent in the direction during January and February. Few easterly directions are observed during the period

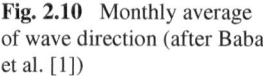

Fig. 2.10 Monthly average
of wave direction (after Baba
et al. [1])

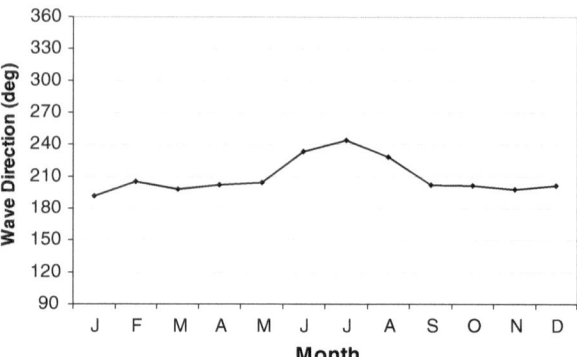

November–January. During the other months from March to October the directions
vary from 152 to 281°N. The westerly components are dominant during the southwest monsoon period of June–August, the range being 227–246°N. During the other
months the mean wave directions are more persistent in the range of 191–205°N.

2.4 Currents

2.4.1 In the Open Sea

For studying the offshore current, the data available from the NIOT DS2
(http://www.niot.res.in) buoy was used. The rose plots show the monthly variation of offshore current for the months of January–May and October–December
(Fig. 2.11). The mean direction of offshore current is predominantly towards
southwest during the months of January, February, March and October. The current speeds vary between 0 and 0.7 m/s with an average value of around 0.2 m/s
during March–May and October–November and around 0.3 m/s during January,
February and December. During the months of April and May the current direction varies between NW and NE. The current direction is mostly towards NW in
November with some scattering in the northerly and westerly directions. NE direction is also equally dominant during December.

2.4.2 Inside Lagoon

Currents near the entrance channel was measured during the period February–
March, 2008. The measured current and its progressive vector plots are presented
in Fig. 2.12. The progressive vector plot indicates net southerly movement. The
drift was towards southeast during the first half of the recording period, followed
by south-southwest movement in the subsequent periods.

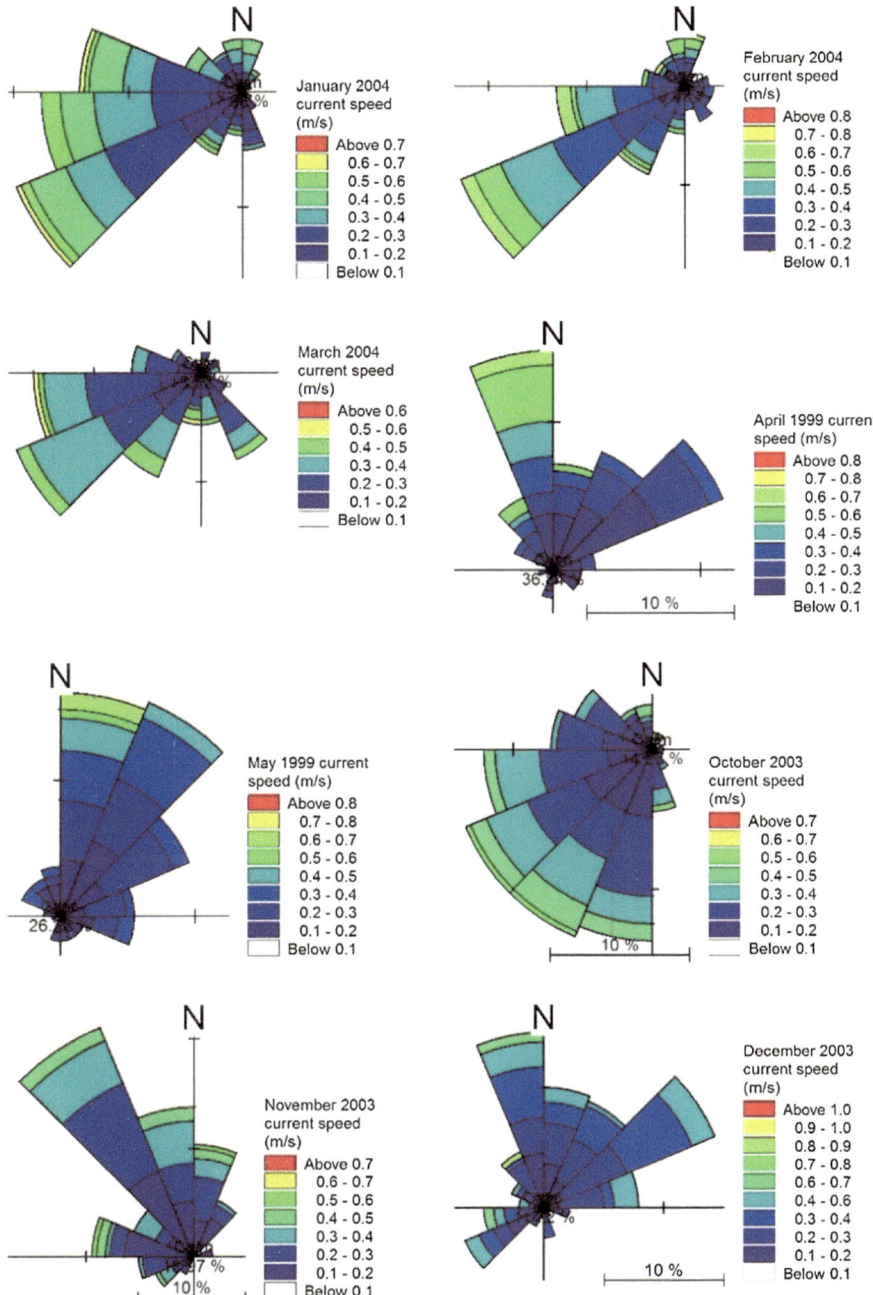

Fig. 2.11 Monthly offshore current rose (*Source* NIOT Buoy-DS2)

Fig. 2.12 Progressive vector plot of current near the entrance channel during February–March, 2008

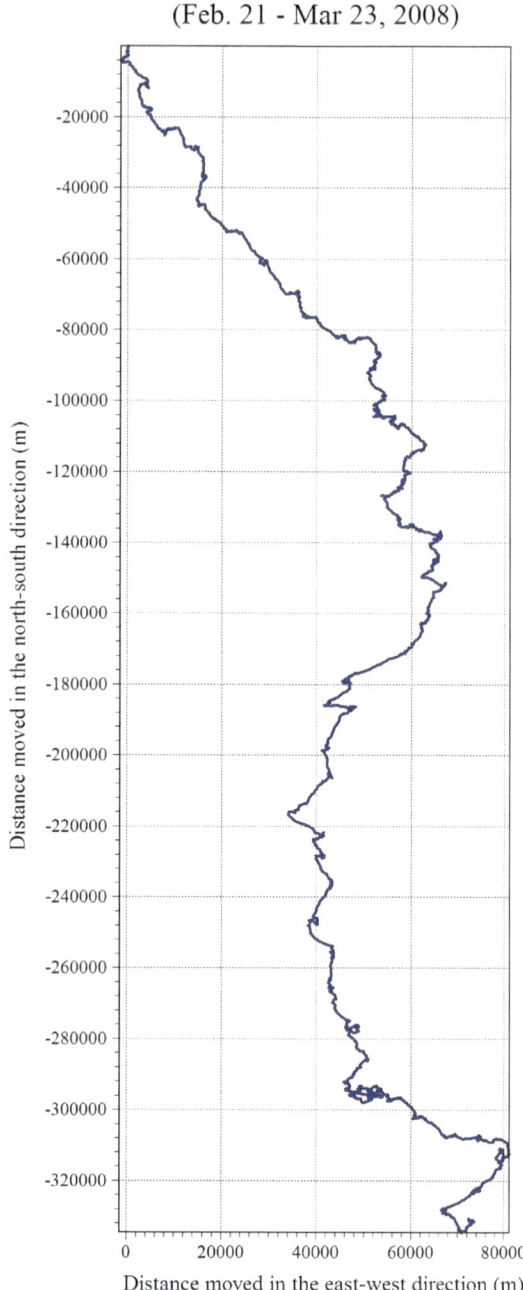

Progressive vector plot
(Feb. 21 - Mar 23, 2008)

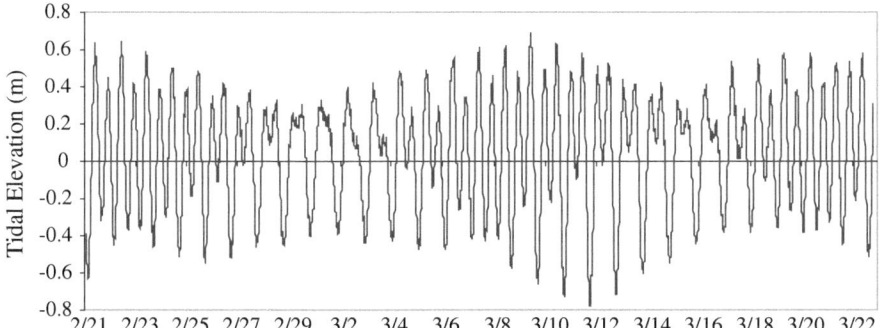

Fig. 2.13 Water level variations inside the lagoon during February–March, 2008

2.5 Tide

Tide is measured using a Wave and Tide Recorder (MIDAS, Valeport, UK) deployed at Kachery jetty of Kavaratti Island at a water depth of 2.5 m [5]. The tides are of the mixed semidiurnal type with a maximum tidal range of 1.4 m. Figure 2.13 shows the water level variation during the period 21st February to 23rd March 2008. As seen in the figure, the tide is found to form a kind of groupings, as seen in the case of waves, with fortnightly lunar cycle, which is indicative of the influence of the constituent MF in the tide.

2.6 Wind

The offshore data from the NIOT buoy DS2 deployed approximately 25 km to the northwest of the Kavaratti Island have been used in this analysis. Figure 2.14 gives the wind rose plots showing the variation of wind during each month. The wind direction is predominantly north-westerly followed by westerly, northerly and northeasterly. During February–March and November–December period, the presence of winds from northeasterly direction is noticeable. During the period January–April and December the wind speed varies between 0 and 8 m/s with a monthly average value of less than 4 m/s with February having the minimum average value of 2.3 m/s. The wind is moderate during the period May–September with average speeds in the range 5.5–8.3 m/s with the minimum in May and the maximum in July. In the month of May the wind speed ranges up to 18 m/s whereas it is up to 14 m/s for June–November.

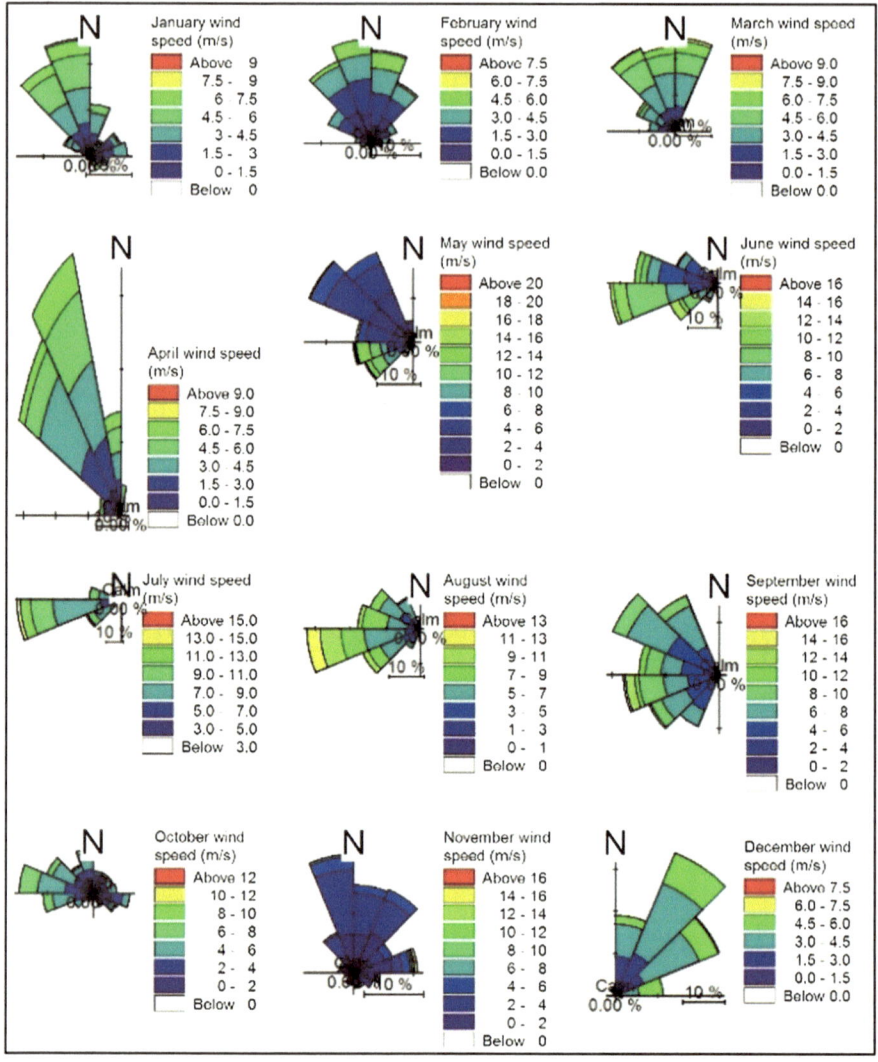

Fig. 2.14 Offshore wind data rose (*Source* NIOT Buoy-DS2)

2.7 Summary

The wave, current, tide and wind characteristics are interlinked in one way or other. The wave characteristics in the coastal waters of the Lakshadweep is derived from the data collected round the year with a directional wave rider buoy. The data reveal that the wave climate of the Lakshadweep Sea is influenced by the south-west monsoon. The period from June to August is the roughest season.

The highest wave height observed during the period is 8.95 m in August. Waves generally do not exceed the height of 5 m during November–March. During the southwest monsoon the dominant values of H_{max} is around 5 m and during the non-monsoon season it is around 1.4 m. Generally H_s are higher during June, July and August when the range is 1.75–4.70 m and are lower during November–March.

The zero-crossing periods (T_z) range from 3.5 to 13.3 s with maximum in February and April and the lowest in June–August. The dominant T_z during the southwest monsoon is in the range of 7–8 s and during the non-monsoon it is 5–7 s. The peak periods (T_p) range from 8.4 to 26.0 s during the year, the lowest being observed in November and the highest in February. Generally, T_p is lower during June–August, as in the case of T_z.

The wave directions ranged from 106 to 316°N during the year. S-SW directions persist throughout the year and these directions dominate during most of the periods except when the southwest monsoon is intense with dominance of westerly waves. Easterly components are observed during November–January.

The mean current direction in the offshore is towards southwest during the measurement periods of January, February, March and October while it is north-northeasterly in April and December. The current speeds ranged up to 0.7 m/s. Net current near the entrance channel is southerly to the lagoon during the month of measurements. The drift is towards southeast during the first fortnight, followed by southwest movement in the subsequent fortnight.

Tide in the Lakshadweep nearshore waters is of the mixed semidiurnal type, with a maximum tidal range of 1.4 m. The tide is fortnightly lunar, which is indicative of the influence of the constituent MF in the tide.

The wind direction is predominantly north-westerly followed by westerly, northerly and northeasterly. During the period January–April and December the wind speed varies between 0 and 8 m/s with a monthly average value of less than 4 m/s. The wind has moderate speed during the months of May, June, July, August and September with average speeds of 5.5–8.3 m/s.

References

1. Baba M, Shahul Hameed TS, Kurian NP, Subhaschandran KS (1992) Wave power of Lakshadweep Islands. Report submitted to Department of Ocean Development, Government of India, Centre for Earth Science Studies, Trivandrum
2. Shahul Hameed TS, Kurian NP, Baba M (1994) Wave climate and power off Kavaratti, Lakshadweep. In: Proceedings of Indian national dock harbour and ocean engineering conference, CWPRS, Pune, pp A63–A72
3. Baba M, Shahul Hameed TS, Kurian NP (2001) Wave climate and wave power potential of Lakshadweep Islands. Geol Surv India Spec Publ 56:211–219
4. Baba M, Thomas KV, Shahul Hameed TS, Kurian NP, Rachel CE, Abraham S, Ramesh Kumar M (1987) Wave climate and power off Trivandrum. In: Project report on sea trial of a 150 kW wave energy device off Trivandrum coast. IIT, Madras on behalf of Dept of Ocean Development, Government of India
5. Prakash TN, Sheela LN, Shahul Hameed TS, Thomas KV, Kurian NP (2010) Studies on shore protection measures for Lakshadweep Islands. Final report submitted to DST, UT Lakshadweep, Centre for Earth Science Studies, Trivandrum

Chapter 3
Beach Morphology

Abstract Baseline data on erosion and accretion in the Laccadive, Amindivi and Minicoy group of islands which were generated during the period 1993–2005, are documented in this chapter. The beach morphological data shows that coastal erosion is a serious problem faced by the islands. Often these changes badly affect the livelihood of the local population, especially during the monsoon season. Erosion in the island is part of a cyclic process during which the beach material is carried away by wave action, tidal currents, littoral currents and other coastal processes. The most serious incidents of coastal erosion in the island were reported during a cyclone/storm event. The studies on coastal erosion would help to make strategies for disaster mitigation measures in the islands.

Keywords Coastal erosion · Beach profiles · Beach volume changes · Short-term and long-term changes · Natural hazards

3.1 Introduction

The coastlines of the Lakshadweep Islands are fully exposed to the hydrodynamic forces resulting in both long-term and short-term variations in shoreline which is pre-dominantly influenced by monsoon. Often these changes can affect the coastal equilibrium leading to coastal erosion at certain locations due to varying hydrodynamic conditions. For a stable beach there is always a balance maintained between the seasonal erosion and accretion without any appreciable loss or gain of material. However, most of the coastal locations face the prospect of erosion or accretion of varying scales due to a combination of natural and anthropogenic factors. The most serious incidents of coastal erosion occur during a storm event resulting in severe loss of beach. In order to develop shoreline management plan for the islands there is a need to conduct site specific beach monitoring programmes coupled with routine collection of environmental data in respect of waves, currents,

© The Author(s) 2015
T.N. Prakash et al., *Geomorphology and Physical Oceanography of the Lakshadweep Coral Islands in the Indian Ocean*, SpringerBriefs in Earth Sciences,
DOI 10.1007/978-3-319-12367-7_3

tides, nearshore bathymetry, sediment properties, etc. As part of this initiative Centre for Earth Science Studies during the period 1990–2005 had undertaken studies for the assessment of both short and long-term shoreline changes in all the islands of Lakshadweep, viz.,—Kavaratti, Agatti, Amini and Bangaram during 1990–1993 [1]; Kadamat, Kiltan, Chetlat and Bitra during 1997–2000 [2] and Androth, Kalpeni and Minicoy during 2003–2005 [3]. The salient observations that emanated from these studies are presented in this chapter which gives an overall picture of the beach morphological changes of the inhabited islands of Lakshadweep.

3.2 Beaches of Lakshadweep

The Island's beaches are made up of coral sand, boulder and pebbles. Sandy beaches are more prevalent on the lagoon coast, while storm beaches made up of coral debris/pebbles are present on the non-lagoon coast. Generally, the sediments of the lagoon coast are finer compared to the non-lagoon side. The sediments are of well sorted to moderately well sorted category with some areas showing even very well sorted category. Beach rocks are common along the intertidal regions of the coast. The islands coasts are micro-tidal with a tidal range of about 1 m. Recreational facility provided by sandy beaches is one of the major tourist attractions of the islands. Loss of functional beaches due to coastal erosion/coastal protection structures are adversely affecting the beach ecosystem and its aesthetics. In the Lakshadweep Islands there are shore parallel natural dunes and also man-made sand dump in the interior part of the island. Good example of natural dunes system can be seen at Amini. Anthropogenically altered dunes are seen at Kavaratti, which are in the interior of the island. In general these sand dumps have heights from 3 to 6 m above the MSL with dune rising to a maximum height of about 6 m on the northeast coast of Amini Island.

3.3 Beach Monitoring Programme

As part of the beach monitoring programme in the inhabited islands, regular beach profiling was carried out from permanent reference stations established based on the coastal morphological signatures at varying intervals between 200 and 500 m along the perimeter of each island. Foreshore sediment sampling and Littoral Environmental Observations (LEO) were also carried out at each station. Totally 5 sets of beach profiles were collected from each island over a period of 3 years representing pre-monsoon and post-monsoon period except Androth, Kalpeni and Minicoy from where only 3 sets were collected. The beach profiles collected during these periods were compared and beach volume changes computed to classify the beaches into eroding, accreting and stable. The sediment samples collected were also subjected to textural analysis to compute the spatial and temporal variations in grain size along the beaches. Data pertaining to the coastal geomorphological features like berm, wave-cut terraces, foreshore slope, etc., were also collected.

3.4 Beach Morphological Changes

In the coastal areas the beach morphological changes are usually reflected in the changes in shoreline position. The shoreline changes can be divided into short-term and long-term. The difference between these two types is only on the time scale. Short-term changes are seasonal and may not lead to a net annual change. In decision making for beach protection measures, the long-term changes are of particular importance.

3.4.1 Long-term Changes

To assess the erosion/accretion trend in the islands, the long-term shoreline changes over a period over 30 years were calculated by comparing the island boundaries of 1967 (from cadastral maps) and High Tide Level (HTL) positions mapped during 1999. The long-term changes of individual islands are presented in Figs. 3.1a–d, 3.2a–e, 3.3 and Table 3.1.

3.4.1.1 Laccadive Group

From Fig. 3.1a–d presenting long term shoreline change trend for the Laccadive group of islands it can be observed that about 36 % (4.15 km) of the shoreline of Kavaratti along the stations CSK 3–8 on the north-west part and stations CSK 13 and 14 along the lagoon coast is affected by erosion. The station CSK 18 also shows eroding tendency.

The percentages of shoreline observed to be eroding are 56, 42 and 21 for the islands of Agatti, Androth and Kalpeni respectively. On the whole erosion along a stretch of the coast in an island is more or less compensated by accretion at another place in the same island. At Agatti Island erosion is noticed on the eastern coast along a stretch of 1.7 km towards north of the airport/helipad. The islands of Androth and Kalpeni have stable shoreline for about 5.2 and 7.31 km of the coast respectively. They are more stable compared to Kavaratti and Agatti (<1 km).

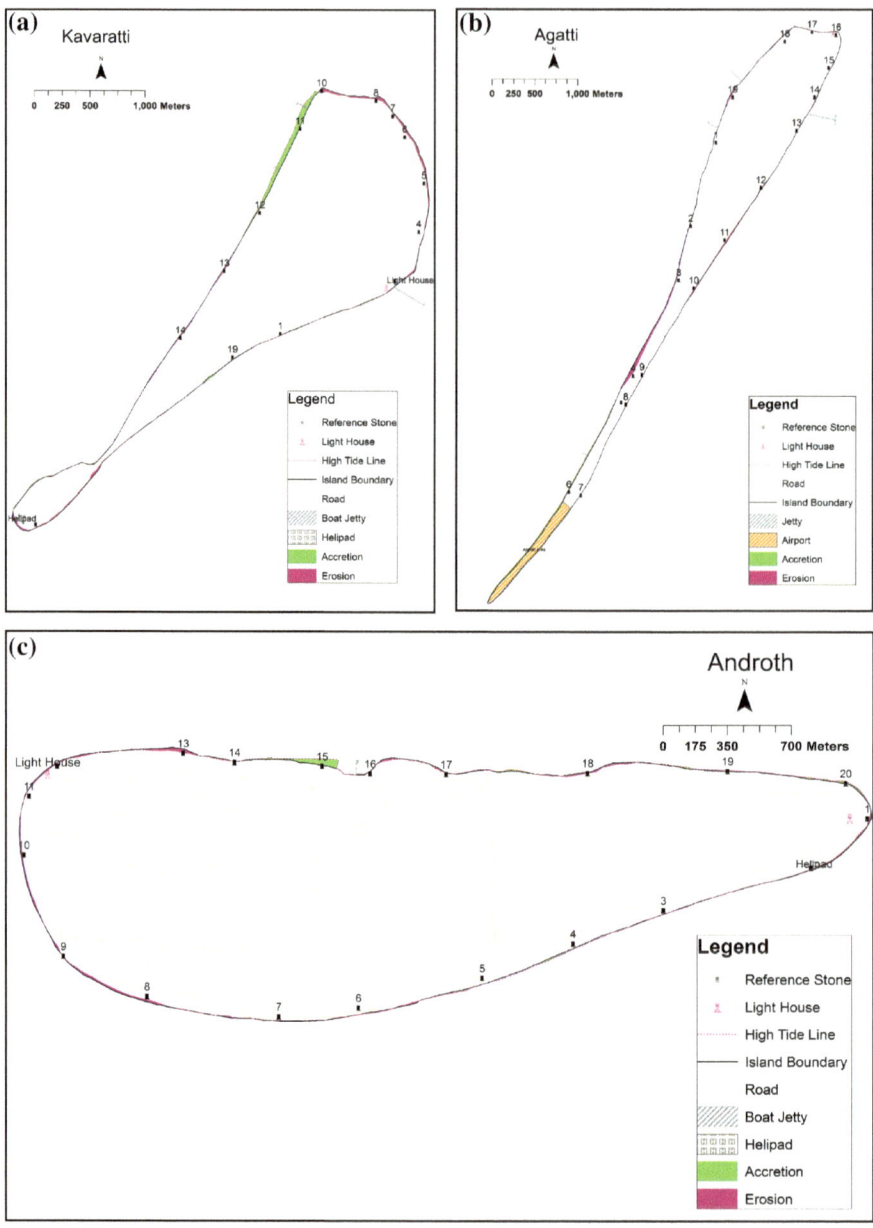

Fig. 3.1 Trend of long-term shoreline changes (1967–1999) for the Laccadive group of islands **a** Kavaratti Island **b** Agatti Island **c** Androth Island **d** Kalpeni Isalnd

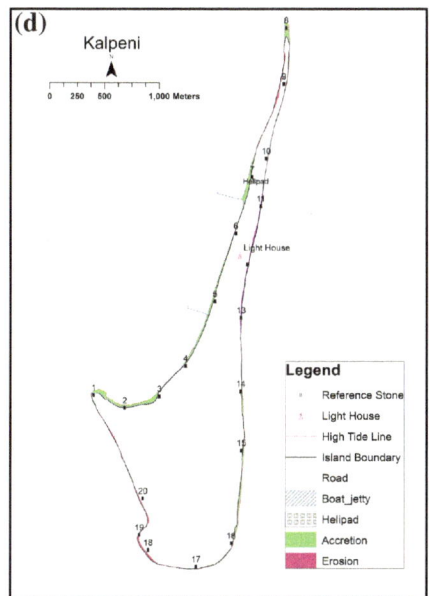

Fig. 3.1 (continued)

3.4.1.2 Amindivi Group

The length of shoreline observed to be eroding in the islands of Amini, Kadmat, Chetlat, Kiltan and Bitra are 2.45, 5.55, 3.64, 2.14 and 0.11 km respectively (see Table 3.1). Among the islands of this group, the maximum shoreline recession of 35 m is observed at Chetlat on the lagoon side of the coast followed by Kiltan with 28 m on the eastern side near the station CSKL 8 (Fig. 3.2a, b). At Amini and Kadmat the maximum recession observed is 20 m each (Fig. 3.1c, d). The majority of the stations at Bitra shows accretion trend whereas at other islands both erosion and accretion are seen (Fig. 3.2e). The long-term shoreline changes show that almost all the islands in this group exhibit an overall accreting trend.

3.4.1.3 Minicoy Group

Minicoy, the only inhabited island in the southern most group shows an overall eroding trend (Fig. 3.3a). The maximum recession of shoreline of 47 m is observed near the jetty. Nearly 10 km of the shoreline which is about 43 % is under erosion whereas the accreting part is very less (3.58 km) compared to the other group of islands. The rest of the shoreline (9.51 km) is found to be stable.

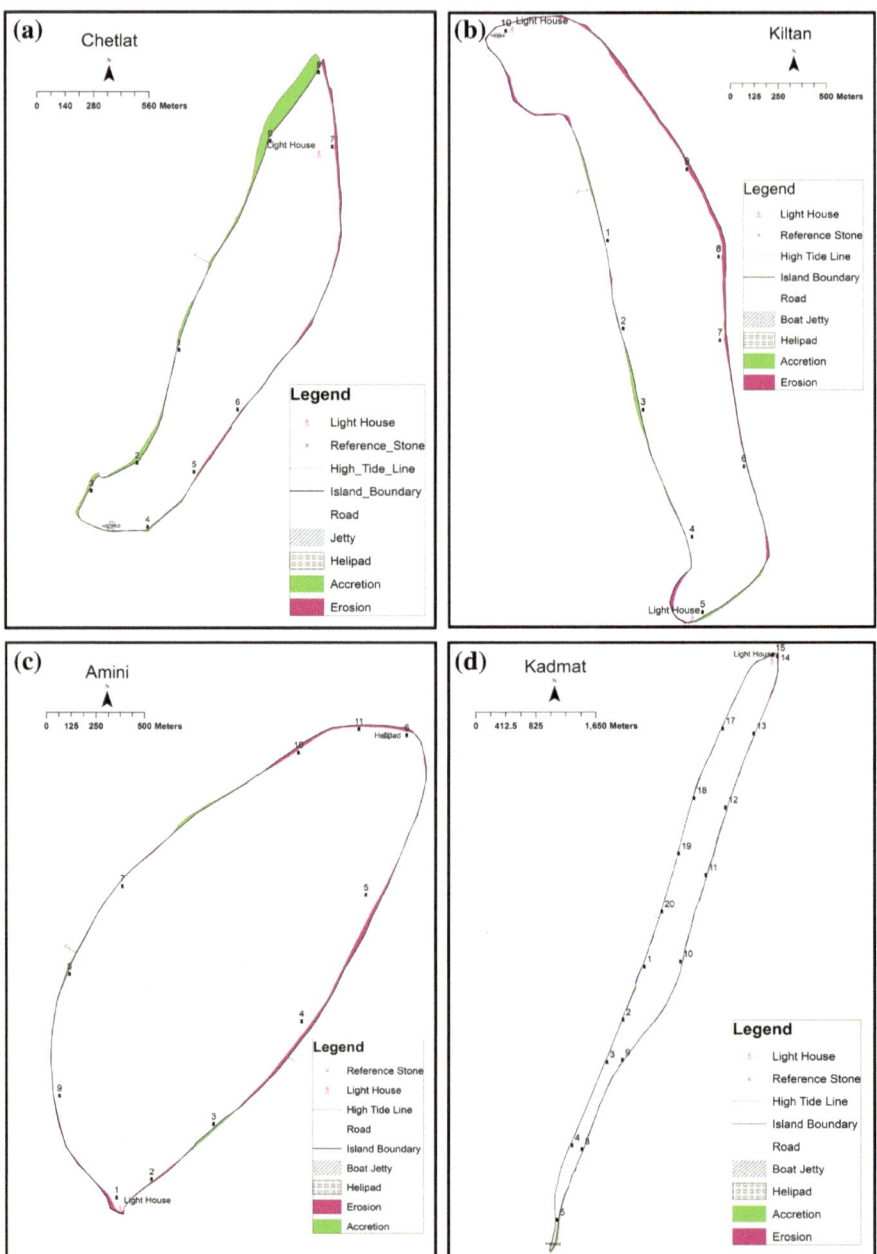

Fig. 3.2 Trend of long-term shoreline changes 1967–1999 for the Amindivi group of islands
a Chetlat Island **b** Kiltan Island **c** Amini Island **d** Kadmat Island **e** Bitra Island

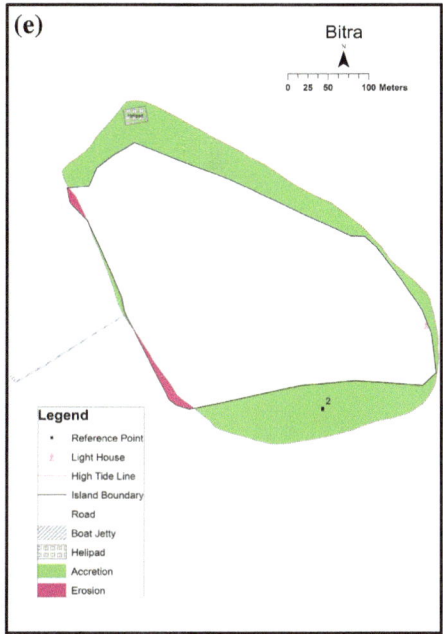

Fig. 3.2 (continued)

3.4.2 Short-term Changes

The short-term variations in shoreline of all the islands during different seasons were studied in detail by measuring beach profiles and is given in Table 3.2. Typical beach profiles representing two distinct locations along the lagoon coast and open coast for selected islands are given in Fig. 3.4. The profiles were compared and beach volume changes computed for each station to assess the stability of the beach at each of the reference station. The data on the cumulative volume changes are used further to classify the beaches into stable, accreting and eroding. The beaches are considered to be stable where the beach volume changes are less than 1 m³/m of beach. The beaches are termed as moderately eroding or accreting when the volume changes are between ±1 and 5 m³/m and highly eroding or accreting when it is ±5 to 10 m³/m. Above this limit is designated as critical. The net volume changes for all the inhabited islands are also presented below.

3.4.2.1 Laccadive Group

Kavaratti: In Kavaratti Island 19 beach monitoring stations (CSK 1–19) at an approximate interval of 250 m were established (see also Fig. 3.1a). The lagoon side of the coast is represented by stations CSK 10–16 whereas stations CSK 1–6 and 17–19 are on the open coast. Stations CSK 7–9 lie along the northern part of

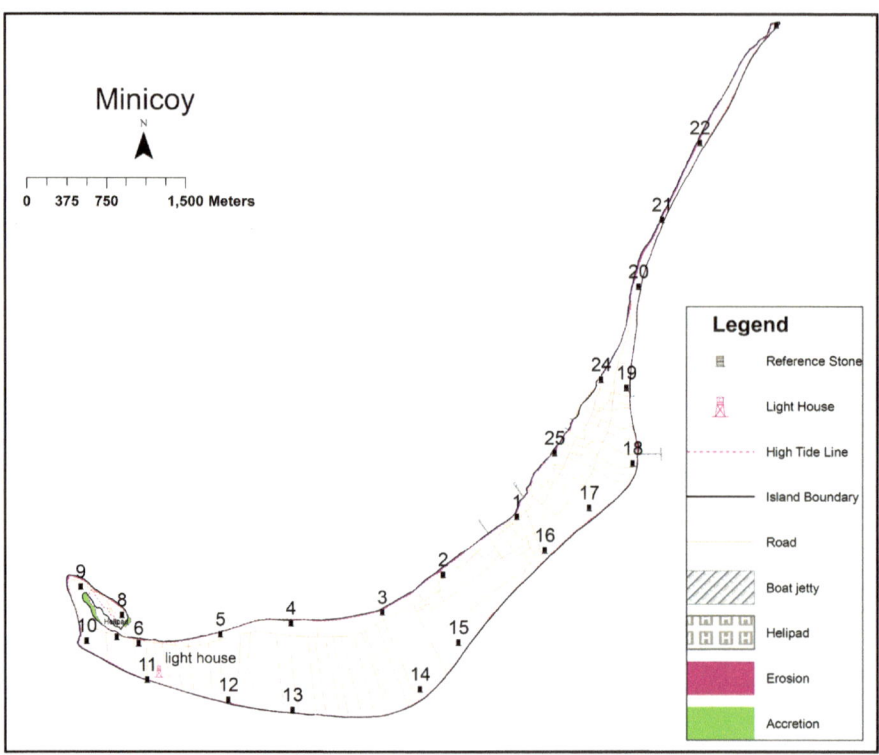

Fig. 3.3 Trend of long-term shoreline changes (1967–1999) for Minicoy Island

Table 3.1 Long-term trend in the retreat/advance of shoreline for the period 1967–1999 [2]

Group	Island	Perimeter (km)	Length of shoreline (km) (% given in brackets)		
			Retreat	Advance	No change
Laccadive	Kavaratti	11.45	4.15 (36)	7.12 (62)	0.18 (2)
	Agatti	16.14	9.01 (56)	6.34 (39)	0.79 (5)
	Androth	10.59	4.47 (42)	0.92 (9)	5.20 (49)
	Kalpeni	11.85	2.53 (21)	2.01 (17)	7.31 (62)
Amindivi	Amini	6.67	2.45 (37)	3.85 (57)	0.37 (6)
	Kadmat	18.37	5.55 (30)	9.82 (54)	3.01 (16)
	Kiltan	7.81	3.64 (47)	3.18 (41)	0.99 (12)
	Chetlat	5.82	2.14 (37)	3.20 (55)	0.48 (8)
	Bitra	1.30	0.11 (9)	1.14 (88)	0.15 (3)
Minicoy	Minicoy	23.07	9.98 (43)	3.58 (16)	9.51 (41)

the island. Totally 5 sets of beach profile surveys were carried out during 1990–1992. The computed net beach volume changes during the study period (March 1990–May 1992) are presented in Table 3.3 and Fig. 3.5.

Table 3.2 Details of beach profile measurements in the islands

Island	Date of measurement
Kavaratti, Agatti, Amini and Bangaram	March, 1990, September 1990, February 1991, September 1999 and May 1992 (1990–1992)
Kavaratti	December 2007 and February 2008
Chetlat, Kiltan, Kadamat and Bitra	April 1997, November 1997, March 1998, November 1998, April 1999
Androth	January, August and November 2003
Kalpeni	January 2003 and November 2003
Minicoy	February 2003 and December 2003

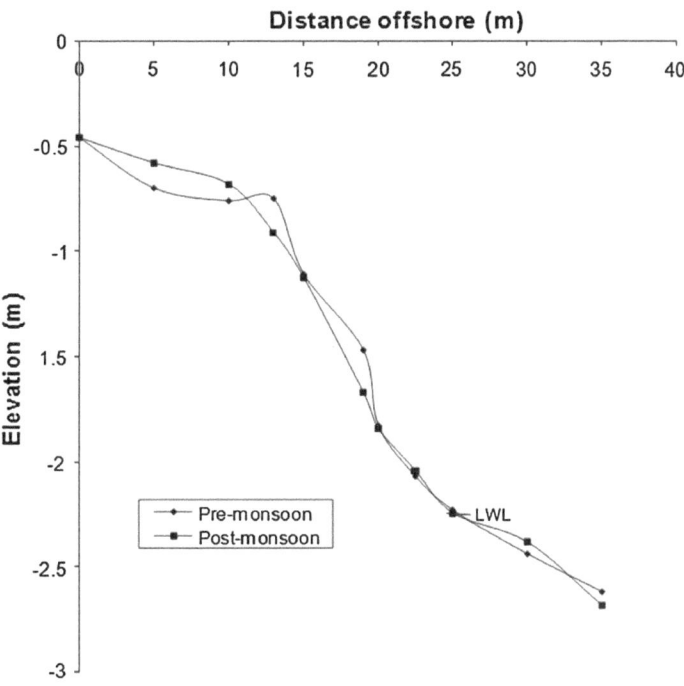

Fig. 3.4 A typical beach profile of the lagoon coast

Majority of the stations in the island show accreting trend except six stations; CSK 4–6, 9 and 15 that show moderately eroding and CSK 17 that shows critically eroding trend. Wave cut terraces of 1–2 m height extending over a length of 100 m are noticed along the beach bordering stations CSK 14 and 15. Along the open coast, CSK 17 on the southeast is critically eroding with a net volume change of -30 m^3/m near the chicken neck, whereas stations CSK 4–6, towards northeast part of the island, experienced moderate erosion during the study period. Remaining stations are found to be accreting. Beach profiles of stations CSK 7–9

Table 3.3 Beach volume change (m³/m) in Kavaratti during March 1990–May 1992

Station	Volume change (m³/m)	
	Accretion	Erosion
CSK 1	6.20	
CSK 2	30.00	
CSK 3	17.10	
CSK 4		−4.15
CSK 5		−2.20
CSK 6		−5.10
CSK 7	4.10	
CSK 8	15.05	
CSK 9		−2.30
CSK 10	4.15	
CSK 11	8.10	
CSK 12	8.30	
CSK 13	4.10	
CSK 14	4.40	
CSK 15		−1.20
CSK 16	4.45	
CSK 17		−30.10
CSK 18	5.10	
CSK 19	2.10	
Total	113.15	−45.05
Net change	68.10	

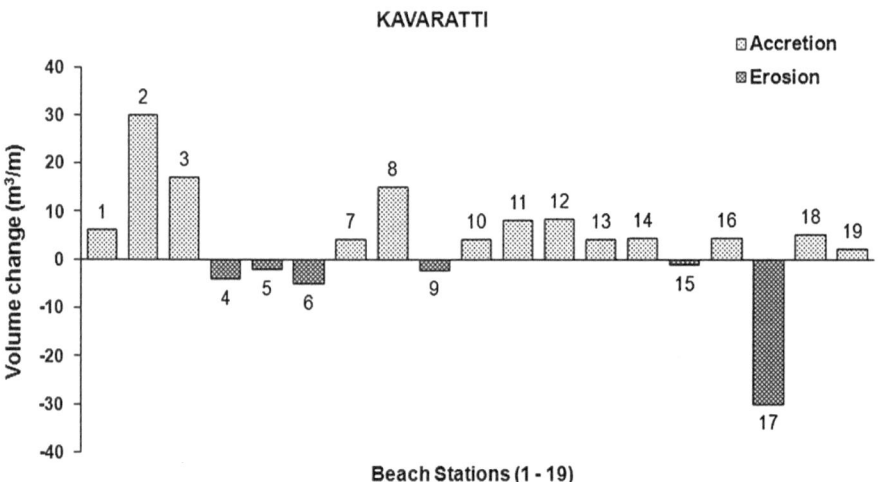

Fig. 3.5 Net beach volume changes for the period of March 1990–May 1992 in Kavaratti

which are on the northern side show signatures of seasonal movement of sediment between the lagoon coast on the west and the open coast on the east. The observed short term trends agree with long term changes in the island.

Agatti: In Agatti Island 19 stations, named as CST 1–19, were established of which 9 stations (CST 7–15) were along the open coast, 2 (CST 16 and 17) on the northern side and remaining 8 (CST 1–6 and 18–19) on the lagoon side (see also Fig. 3.1b). The net volume changes computed for the period March 1990–May 1992 are shown in Table 3.4 and Fig. 3.6.

Most of the stations located along the lagoon coast experienced erosion during the study period. Continuous erosion was noticed at CST 2–6 wherein the station CST 5 was critically eroding with a net volume change of nearing 16 m³/m. To combat the high rates of erosion various shore protection measures like placing of hollow concrete blocks and tetrapods were adopted in this area. Moderate accretion is noticed at station CST 1, located to the south of main jetty, whereas CST 19 which lies between the two jetties on the lagoon side experiences moderate erosion and this can be linked to the hindrance in sediment transport due to

Table 3.4 Beach volume change (m³/m) in Agatti during March 1990–May 1992

Station	Volume change (m³/m)	
	Accretion	Erosion
CST 1	10.1	
CST 2		−3.52
CST 3		−8.30
CST 4		−1.30
CST 5		−15.60
CST 6		−6.80
CST 7		−13.60
CST 8	2.80	
CST 9	12.10	
CST 10		−8.50
CST 11	6.20	
CST 12	5.00	
CST 13	7.80	
CST 14	9.10	
CST 15	0.60	
CST 16	21.15	
CST 17	4.80	
CST 18	8.10	
CST 19		−6.10
Total	87.75	−63.72
Net change	24.03	

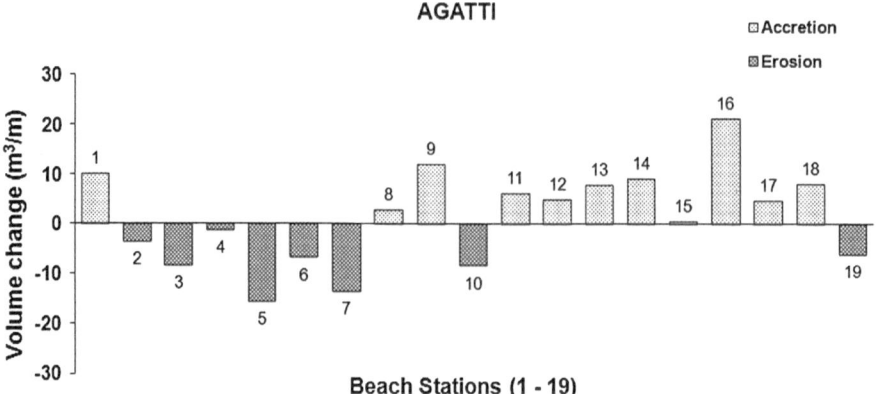

Fig. 3.6 Net beach volume changes (m³/m) in Agatti for the period March 1990–May 1992

the construction of structures on either side. Along the open coast station CST 7 located on the southeast adjacent to the airport showed critical erosion whereas CST 10 located further north was highly eroding. The stations CST 17 and 18 on the northern part of the island show moderate to high accretion and this can be attributed to the movement of sediment from the lagoon coast on the western side to the open coast on the east.

Androth: Androth is the biggest island among the inhabited islands (Fig. 3.1c, also see Table 1.1). In this island, 20 beach monitoring stations (CSA 1–20) were established, approximately, at 500 m intervals. Out of these, 8 are on southern coast (CSA 1–8), 4 on the western side (CSA 9–12) and remaining on the northern part of the island (CSA 13–20). The erosion/accretion rates are given in Table 3.5 and Fig. 3.7.

Majority of the stations along the southern part of the island except CSA 1 and 7 show erosion. High erosion is observed at CSA 3 and 5 with values of 153.66 and 84.5 m³/m respectively. Further along the southwestern part of the island the beaches are moderately eroding with values of 10.6 and 7.6 m³/m at CSA 9 and CSA 10 respectively. The beaches on the northern part of the island up to the breakwater are more or less stable except at stations CSA 12 and 14, which are moderately eroding. The beach at station CSA 16 inside the harbour is stable. This indicates that the western arm of the breakwater has prevented free flow of alongshore sediment. Erosion is observed at stations further east of the break water with station CSA 18 showing an erosion of 14 m³/m. However the station CSA 20, on the eastern most end of the island shows a high accretion of 31.1 m³/m.

Kalpeni: In Kalpeni Island, 20 beach monitoring stations were established, along the coast approximately, at 300 m intervals (see also Fig. 3.1d). The stations CSK 1–8 and 18–20 represent the lagoon coast whereas CSK 9–17 are on the open coast. The net beach volume change for the study period

Station	Volume change (m^3/m)	
	Accretion	Erosion
CSA 1	11.46	
CSA 2		−32.38
CSA 3		−153.66
CSA 4		−21.32
CSA 5		−84.5
CSA 6		−24.78
CSA 7	5.58	
CSA 8		−3.41
CSA 9		−10.63
CSA 10		−7.55
CSA 11	1.59	
CSA 12		−6.85
CSA 13		−2.03
CSA 14		−6.69
CSA 15	7.84	
CSA 16		−0.45
CSA 17		−0.41
CSA 18		−13.97
CSA 19		−5.36
CSA 20	31.1	
Total	57.57	−377.4
Net change	−319.43	

Table 3.5 Beach volume change (m^3/m) in Androth during January 2003–November 2003

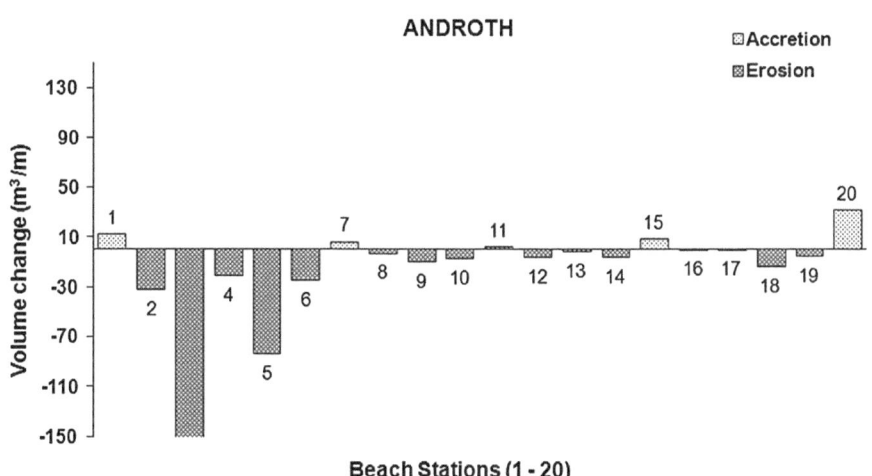

Fig. 3.7 Net beach volume changes during January 2003–November 2003 in Androth

Table 3.6 Beach volume change (m^3/m) in Kalpeni during January 2003–November 2003

Station	Volume change (m^3/m)	
	Accretion	Erosion
CSK 1		−1.75
CSK 2		−33.88
CSK 3	1.87	
CSK 4		−3.85
CSK 5		−32.59
CSK 6		−0.15
CSK 7		−8.40
CSK 8 lagoon		−5.05
CSK 8 sea		−8.40
CSK 9 lagoon		−7.85
CSK 9 sea		−2.96
CSK 10		−6.05
CSK 11	29.21	
CSK 12		−1.01
CSK 13		−10.46
CSK 14		−1.86
CSK 15		−3.18
CSK 16		−2.19
CSK 17	3.95	
CSK 18		−2.45
CSK 19		−4.03
CSK 20		−66.94
Total	31.10	−199.10
Net change	−168	

January 2003–November 2003 are given in Table 3.6 and Fig. 3.8. Since the study period is not confined to the full year, the volume change indicates the seasonal trend rather than annual.

A net erosion of 68 m^3/m is observed for the period all along the coast (Table 3.6). Majority of the stations except CSK 11 and CSK 17 show an eroding tendency because of the monsoonal impact.

3.4.2.2 Amindivi Group

Amini: A total of 16 coastal stations (CSA 1–16) were established for beach monitoring at this island. Out of this 8 stations (CSA 1–4 and 13–16) are located on the lagoon side, one each (CSA 5 and CSA 12) on the northern and southern tip and

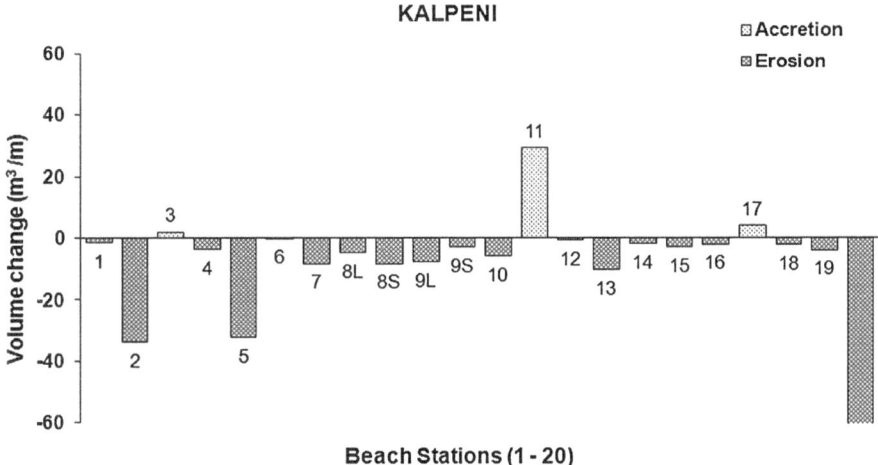

Fig. 3.8 Net beach volume changes during January 2003–November 2003 in Kalpeni

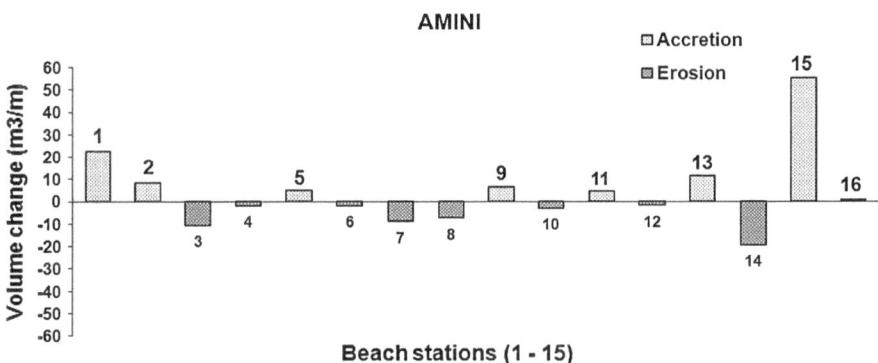

Fig. 3.9 Net beach volume changes for Amini Island during March 1990–May 1992

the remaining 6 stations (CSA 6–11) along the open coast on the eastern side (see also Fig. 3.2c). The computed beach volume changes for the period March 1990–May 1992 are presented in Fig. 3.9 and Table 3.7.

This island is notable for an overall net accretion of 57.05 m³/m. Station CSA 15 on the lagoon coast has a high accretion of 55 m³/m followed by adjoining station CSA 1 with an accretion of 22.5 m³/m. A location which is critically eroding in this island is CSA 14 on the lagoon coast towards north of jetty where a net erosion of 19.6 m³/m is observed. This is followed by CSA 3 on the southwestern part of the island with an erosion of 10.8 m³/m. Stations CSA 7 and 8 on the open coast show moderate erosion as indicated by volume changes of 7.5 and 9.0 m³/m respectively.

Table 3.7 Net beach volume change in Amini Island during March 1990–May 1992

Reference station	Volume change (m^3/m)	
	Accretion	Erosion
CSA 1	22.50	
CSA 2	8.20	
CSA 3		−10.80
CSA 4		−2.10
CSA 5	4.80	
CSA 6		−2.00
CSA 7		−9.00
CSA 8		−7.50
CSA 9	6.20	
CSA 10		−3.15
CSA 11	4.6	
CSA 12		−1.80
CSA 13	11.10	
CSA 14		−19.60
CSA 15	55.0	
CSA 16	0.6	
Total	113	−55.95
Net change	57.05	

On the NW part of the island between stations CSA 12 and 13, sand dunes are as high as 6 m above Mean Sea Level (MSL) are seen. Moderate erosion is observed at station CSA 12 on the northern part of the island where wave cut terraces are noticed resulting in slumping of the beach face slope.

Kadmat: Kadamat is the longest island in the group (see also Fig. 3.2d). In this island, 20 beach monitoring stations (named as CSK 1–20) were established, approximately, at 500 m intervals. Out of these, 12 are on the lagoon coast (CSK 15–20 and CSK 1–6) and the remaining 8 on the east coast (CSK 7–14). The beach volume changes for the period April 1997–April 1999 have been calculated and are presented in Table 3.8 and Fig. 3.10.

In the lagoon coast, maximum erosion is observed to the south of the jetty, at stations CSK 1 and 2 with values of 5.2 and 7.3 m^3/m respectively. Further south, the beaches at stations CSK 4–6 are found to be stable or moderately accreting. The beaches on the northern side of the jetty are stable except the station CSK 20 adjoining the jetty which like station CSK 1 show moderate erosion. Majority of the stations on the east coast of the island shows accretion with stations CSK 7 and CSK 8 on the southeast side leading with 15.1 and 8.5 m^3/m respectively. The station CSK 15 on the northwest tip of the island is found to undergo erosion with rates comparable to CSK 1 and 2 south of the jetty. In general there is considerable accretion on the east coast whereas the lagoon coast, especially to the south of the jetty is under erosion.

Chetlat: In Chetlat Island, 12 beach monitoring stations were established, approximately, at 300 m interval (see also Fig. 3.2a). The stations, which represent

Table 3.8 Beach volume change (m^3/m) in Kadmat Island during April 1997–April 1999

Reference station	Volume change (m^3/m)	
	Accretion	Erosion
CSK 1		−5.15
CSK 2		−7.28
CSK 3		−1.54
CSK 4	4.42	
CSK 5		−0.71
CSK 6	5.62	
CSK 7	15.11	
CSK 8	8.51	
CSK 9		−3.26
CSK 10	0.93	
CSK 11	3.37	
CSK 12	1.23	
CSK 13	2.34	
CSK 14	3.38	
CSK 15		−6.61
CSK 16		−0.54
CSK 17	1.84	
CSK 18		−0.46
CSK 19	2.07	
CSK 20		−1.85
Total	+48.82	−27.4
Net change	+21.42	

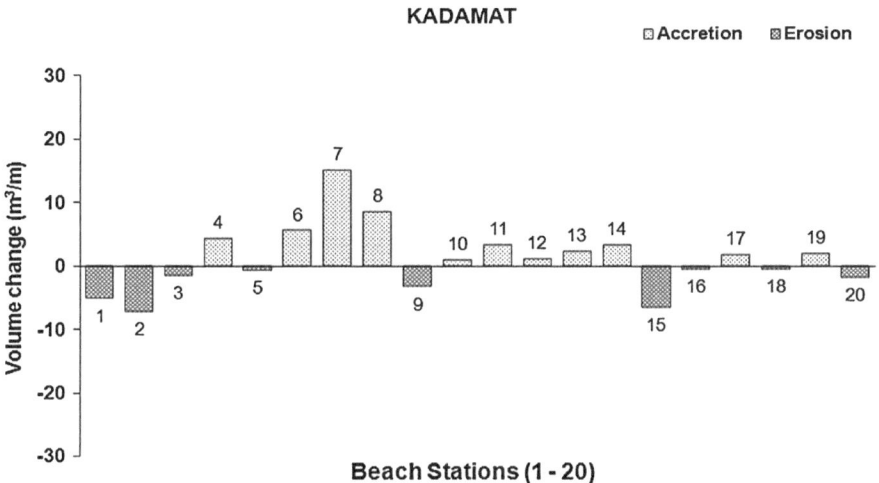

Fig. 3.10 Net beach volume changes for Kadamat Island during April 1997–April 1999

the lagoon coast are CSC 11 to 12 and 1 to 4, and CSC 5 to 9 represents the east coast. Erosion/accretion rates were computed for the period April 1997–April 1999 for all the stations as given in Table 3.9 and Fig. 3.11.

In general accretionary trend is dominating for the lagoon coast, barring station CSC 1 (south of Jetty), which is critically eroding at a rate of 12.80 m^3/m. High accretion (10.03 m^3/m) is noticed at station CSC 3, followed by CSC 11 with a deposition of 8.96 m^3/m. Most of the beaches on the eastern coast are either stable or have a tendency for erosion like station CSK 6. On the northern part of the island at station CSC 10, seasonal transport

Table 3.9 Beach volume change (m^3/m) of Chetlat Island during April 1997–April 1999

Reference station	Net change (m^3/m)	
	Accretion	Erosion
CSC 1		−12.80
CSC 2	1.58	
CSC 3	10.01	
CSC 4	3.14	
CSC 5		−2.31
CSC 6		−4.32
CSC 7	3.75	
CSC 8	0.80	
CSC 9		−3.68
CSC 10	3.72	
CSC 11	8.96	
CSC 12	2.20	
Total	34.16	−23.11
Net cumulative change	+11.05	

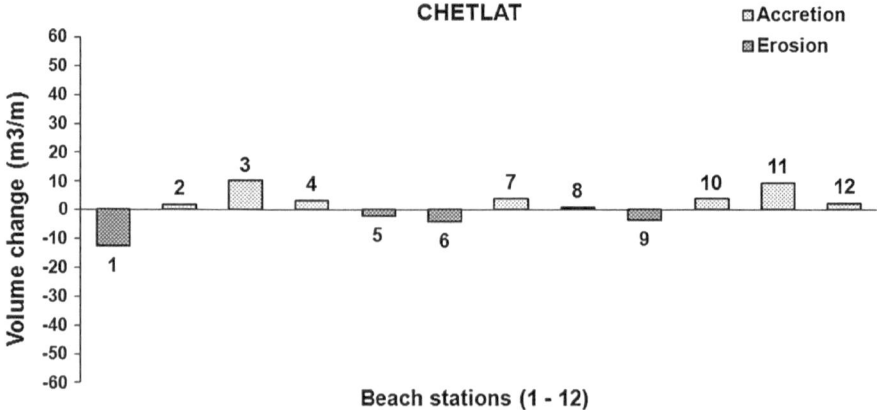

Fig. 3.11 Net beach volume changes for Chetlat Island during April 1997–April 1999

of sediments between the lagoon coast and the eastern part of the island is observed; however, the net change observed at this station is a slight accretion of 3.7 m³/m.

Kiltan: In Kiltan Island, 11 beach monitoring stations (CSKL 1–11) were established (see also Fig 3.2b), of which CSKL 1–5 and 11 were on the lagoon coast, CSKL 10 on the northern tip and CSKL 6–9 on the east coast during the study period of April 1997–April 1999. The erosion/accretion rates were calculated for each station and are presented in Table 3.10 and Fig. 3.12.

In general there is an overall domination of accretion in this island. Stations CSKL 3 and 4 on the lagoon coast show erosion with a volume change of 6 and

	Reference station	Net change (m³/m)	
		Accretion	Erosion
Table 3.10 Beach volume change (m³/m) of Kiltan Island during April 1997– April 1999	CSKL 1	5.72	
	CSKL 2	9.39	
	CSKL 3		−5.99
	CSKL 4		−4.26
	CSKL 5	6.49	
	CSKL 6	0.59	
	CSKL 7	5.44	
	CSKL 8		−4.68
	CSKL 9	0.74	
	CSKL 10		−9.99
	CSKL 11	8.01	
	Total	36.38	−24.92
	Net change	+11.46	

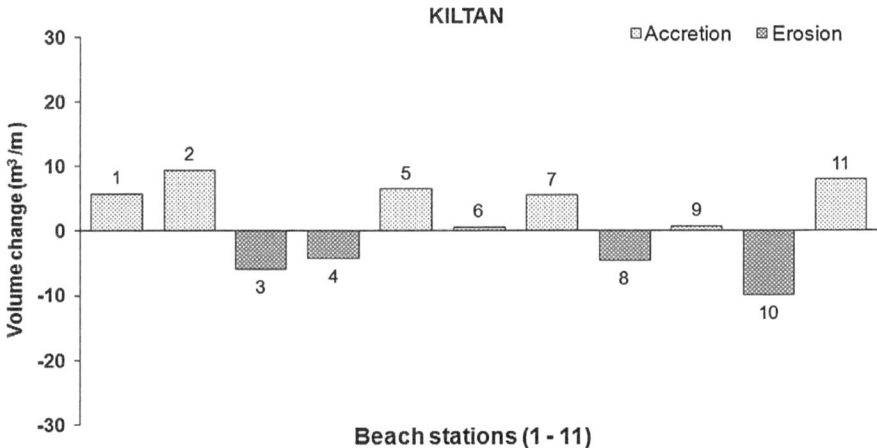

Fig. 3.12 Net beach volume changes for Kiltan Island during April 1997–April 1999

4.3 m³/m respectively. The remaining stations in the lagoon coast are accreting. It can also be noted that station CSKL 1 is well protected by tetrapods. Station CSKL 11, north of jetty shows high accretion of 8 m³/m.

Along the east coast, all stations barring CSKL 8 show accretion. However, high erosion of 10 m³/m is observed at station CSKL 10, which is at the northern tip of the island. As in the case of other islands, considerable seasonal exchange of sediments takes place at this station in between the lagoon coast and east coast. However, unlike the other islands, there is a significant eroding tendency here.

Bitra: The net beach volume change computed for the period March 1990–May 1992 for 8 stations CSB 1–8 is given in Fig. 3.2e. It can be seen that a majority of the stations face erosion. However, the quantum of accretion is much higher at the stations CSB 1 and 2, on the south, while stations CSB 3–6 on the east and CSB 8 on the west showed erosion of various magnitudes (Fig. 3.13).

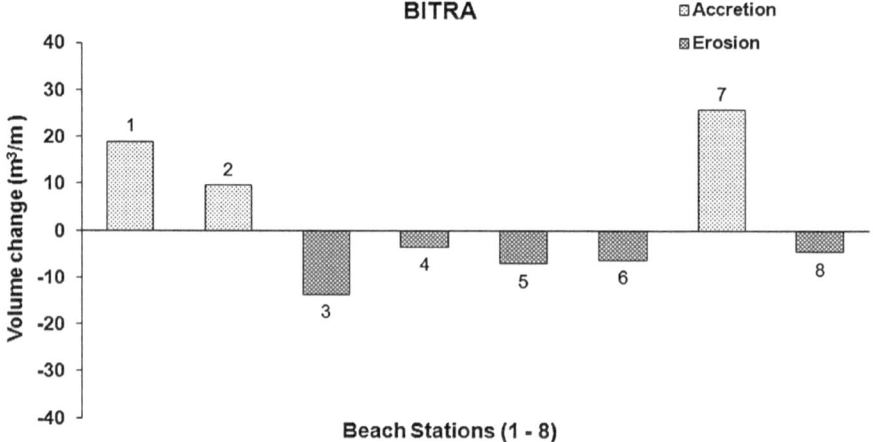

Fig. 3.13 Net beach volume changes for Bitra Island during April 1997–April 1999

Table 3.11 Beach volume change (m³/m) of Bitra during April 1997–April 1999	Reference station	Net change (m³/m)	
		Accretion	Erosion
	CSB 1	18.86	
	CSB 2	9.5	
	CSB 3		−13.65
	CSB 4		−3.45
	CSB 5		−7.06
	CSB 6		−6.29
	CSB 7	25.76	
	CSB 8		−4.5
	Total	54.12	−34.95
	Net change	+19.17	

In the lagoon coast, maximum accretion was observed at station CSB 1 (18.86 m³/m). Station CSB 3 in the southern part of the island showed maximum erosion (13.65 m³/m). Station CSB 7 on the open (eastern) coast, showed the highest accretion of 25.76 m³/m (Table 3.11). It is inferred that the beach to the south of jetty and northern part of the island are showing accretion trend while erosion is mostly observed along the south-eastern and open coast of the island during the study period.

3.4.2.3 Minicoy Group

25 stations named as CSM 1–25 (see also Fig. 3.3) were established in Minicoy Island for beach monitoring during the period January 2003 and November 2003. CSM 1–8, 20–23 and 24–25 represent the lagoon coast (west) while the remaining station viz., CSM 9–19 and 20–23 are along the open coast. The erosion/accretion rates for each station are given in Table 3.12 and Fig. 3.14. Since the data does not pertain to the full year and covers only part of the fair weather beach building period, there is a domination of the monsoon erosive period in the data. CSM 24 on the north and CSM 8 on the southern end of the island facing the lagoon recorded high accretion of 21.8 and 19 m³/m respectively. But stations CSM 4 and 5 along the lagoon coast showed erosion with a high value of 19.11 m³/m at CSM 5. At these stations tetrapods were provided for shore protection. For the north, high erosion is also observed at station CSM 21 on the lagoon coast.

On the southern part of the island the beaches along the open coast, which are mostly composed of coarse sand mixed with pebble/shingles are relatively stable. Along the east coast, majority of the stations show erosion. Maximum erosion is observed at station CSM 20 and 23 (east) located on the northern tip of the island. The beaches on the central part of the island particularly stations CSM 14 and 15 show moderate erosion trend whereas at the other locations the beaches are either stable or moderately accreting.

3.5 Beach Sediment Characteristics

The islands are usually with a lagoon on its western margin and a narrow storm beach on the eastern side. The lagoon side beach comprises of sandy type sediments whereas the sea side beach is made of pebbly/boulder sediments (Fig. 3.15). Outside the reef, there are well-defined wave-cut terraces extending from the reef margin to a distance of more than 50–100 m which indicates the lowest sea level during the last 10,000 years. Various types of processes are acting on these islands. A large number of different and diverse biological communities occur in these atolls, which include species of corals, gastropods, bivalves, foraminifers, echinoderms, etc. Coralline algae are a very important group in building the reef and

Table 3.12 Beach volume change (m^3/m) of Minicoy during February 2003 and December 2003

Station	Volume change (m^3/m)	
	Accretion	Erosion
CSM 1	6.98	
CSM 2		−1.19
CSM 3	0.97	
CSM 4		−3.28
CSM 5		−19.71
CSM 6		−0.02
CSM 7	2.10	
CSM 8	19.71	
CSM 9	0.91	
CSM 10		−5.28
CSM 11		−0.16
CSM 12		−11.95
CSM 13	1.05	
CSM 14		−4.57
CSM 15		−4.45
CSM 16	10.8	
CSM 17	10.2	
CSM 18		−2.55
CSM 19	0.95	
CSM 20 East		−23.93
CSM 20 West	0.24	
CSM 21 East		−5.3
CSM 21 West		−21.35
CSM 22 East	11.07	
CSM 22 West		−0.11
CSM 23 East		−19.11
CSM 23 West		−6.76
CSM 24	21.82	
CSM 25		−0.22
Cumulative volume change	86.80	−129.94
Net volume change	−43.14	

also in producing sediment from the reef. The main source of the sediments in the atolls is the breakage of corals. The sediments in the lagoon consist of different species of coral reefs like halimoda, shells of gastropods, bivalves, foraminifera, ostracods and bryozoans. The sediments in the atolls are pure skeletal carbonate sands with a very low amount of silica, alumina and iron. The grain sizes of sediments of a typical island viz. Kavaratti during the pre-monsoon, monsoon and post-monsoon seasons are presented in Table 3.13.

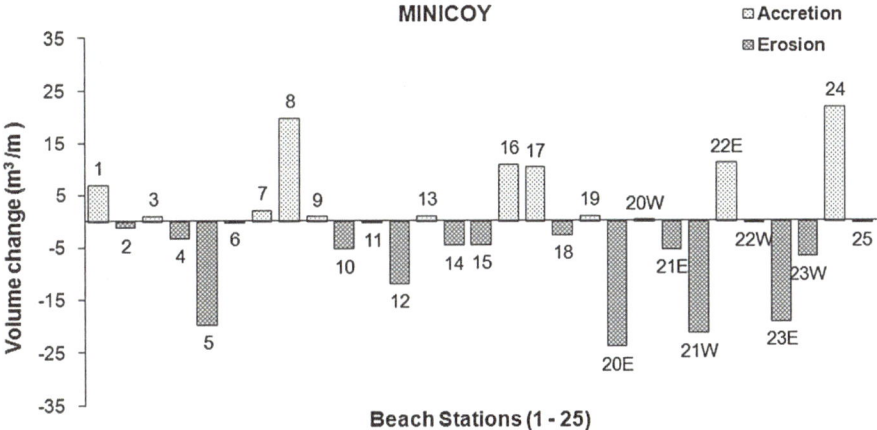

Fig. 3.14 Net beach volume changes for the period of February 2003–December 2003 in Minicoy Island

Fig. 3.15 Pebbly/Boulder beach on the eastern coast of Minicoy Island

3.6 Shore Protection Measures

In order to provide preventive measures to combat coastal erosion the Lakshadweep Administration has implemented many low cost shore protection measures in the island. This is mainly based on the report of a committee contributed by the Union

Table 3.13 Seasonal grain size variation of the foreshore sediments (MWL) at Kavaratti Island

Station	Mean (mm)			Sorting (phi)		
	Pre-monsoon	Monsoon	Post-monsoon	Pre-monsoon	Monsoon	Post-monsoon
1	–	0.61	0.39	–	0.80	0.87
2	0.43	0.48	0.48	0.79	0.78	0.96
3	0.44	0.36	0.41	0.88	0.97	0.77
4	0.45	0.50	–	0.61	0.81	–
5	0.36	0.54	–	0.51	0.85	–
6	0.46	0.68	0.38	0.55	0.75	0.52
7	0.94	0.46	0.32	0.93	1.42	1.12
8	0.47	0.48	0.32	0.93	0.83	0.88
9	0.35	0.50	–	0.70	1.35	–
10	0.21	0.21	0.23	0.56	0.60	0.72
11	0.27	0.27	0.31	0.78	0.68	0.76
12	0.36	0.31	0.35	0.87	0.73	0.84
13	0.38	0.31	0.29	0.76	0.67	0.67
14	0.57	0.50	–	0.48	0.85	–
15	–	0.38	0.48	–	0.85	1.18
16	0.38	0.32	0.67	0.72	0.61	0.89
17	0.63	0.59	0.51	0.56	0.70	0.49
18	–	0.54	0.51	–	0.59	0.57
19	0.38	–	0.40	0.77	–	0.77
20		–	–		–	–

Ministry of Water Resources in 1986 to suggest alternative less costlier measures to prevent erosion in the Lakshadweep Islands [4]. The Committee had suggested a few low-cost pilot schemes for implementation in the islands. Some of the low-cost schemes like hollow concrete blocks, coir bag filled stones, timber piles and planking and tetrapod were adopted by the Public Works Department of Lakshadweep Fig. 3.16a, b. These schemes are found to be less expensive than the conventional rubble seawall and are quite effective in controlling erosion. In location where these structures are constructed it has been observed that the erosion has shifted to the down current side of the coast.

With the high density of population and major establishments along the island coasts, large investments have been made for implementing shore protection structures and to take up other mitigatory measures. While the present protection measures are justified in the above perspective, from the environmental angle the engineering solutions are not always welcome. Maintenance of the beach can be achieved by an appropriate management approach, which will have a judicious mix of different types of interventions including protection measures developed for the particular location through numerical modelling based on hydrodynamic data, near shore bathymetry, sediment properties, etc. These aspects are discussed in Chap. 4.

Fig. 3.16 a Hollow concrete blocks and tetrapods protecting the coast of Minicoy Island.
b Another view of hollow concrete blocks and tetrapods placed on the slope

3.7 Summary

Coastal erosion is one of the recurring natural hazards in Lakshadweep. An analysis of long-term and short-term shoreline change in the different group of Lakshadweep Islands has been carried out by comparing the present High Tide Line (HTL) with island boundary of 1967 and beach profile data. At Kavaratti Island erosion is observed along the SW, SE and NE parts of the island. A critically eroding beach is noticed at station CSK 17 near the chicken neck area where the width of island is less than 50 m. Generally at Agatti Island seasonal erosion is observed along the lagoon coast particularly along the SW and SE part of the island. At Androth Island majority of the stations along the southern part experienced erosion. The presence of breakwater on the northern part of the island is causing hindrance to the free flow of alongshore sediment towards east leading to erosion. At Kalpeni the beaches located to the north of jetty and along the southwest part of the island experienced seasonal erosion.

The Amindivi group of islands which represents the northern group is prone for cyclones and storms. At Amini Island the critically eroding station CSA 14 on the lagoon coast shows a net erosion of 19.6 m^3/m. At Kadamat Island there is considerable accretion on the east coast whereas the lagoon coast, especially to the south of the jetty is under erosion. At Chetlat Island the study has indicated that the beaches to the south of Jetty and SE part shows seasonal erosion. Even at Kiltan Island the beaches towards the south of Jetty shows erosion whereas on the east coast the beaches are stable. Majority of the beaches at Bitra Island are stable in nature except the SE part showing erosion.

At Minicoy Island, beaches on the southeast part which are mostly composed of coarse sand mixed with pebble/shingles are stable compared to the east coast. The beaches on the central part of the island particularly stations CSM 14 and 15 show moderate erosion trend whereas at other locations the beaches are either stable or moderately accreting.

The causes of erosion in the islands are both natural and anthropogenic. While the natural causes are the hydrodynamic processes and reduction in the height of reef edge over a period of time, the anthropogenic factors are construction of coastal protection measures, and beach sand mining. Beach can be maintained through appropriate management plans. In locations where shore protection measures are provided by layers of tetrapods, hollow concrete blocks and coir bag filled with pebbles, the shores are stabilised. Often these kinds of measures are hindrance to the fishing and tourism industry. It has been observed that in some islands there is large scale seasonal transport of sediments from one side of the island to the other; the construction of groins may be effective in controlling erosion. Long term monitoring of shorelines and collection of littoral environmental parameters for each island is recommended. Further, Integrated Coastal Zone Management Plan developed for the islands (Chap. 6) have to be implemented meticulously.

References

1. Suchindan GK, Prakash TN, Prithviraj M (1993) Studies on coastal erosion, sediment movement and bathymetry in selected islands of the UT of Lakshadweep (1990–1993)—Kavaratti, Agatti, Amini and Bangaram (Phase-I). Technical report, Department of Science and Technology, UT Lakshadweep, CESS, Trivandrum
2. Prakash TN, Suchindan GK, Prithviraj M (2001) Studies on coastal erosion studies in Lakshadweep Islands (1997–2000)—Kadamat, Chetlat, Kiltan and Bitra (Phase-II). Technical report, Department of Science and Technology, UT Lakshadweep
3. Prakash TN, Suchindan GK, Thomas KV (2005) Studies on coastal erosion studies in Lakshadweep Islands (2002–2005)—Androth, Kalpeni and Minicoy (Phase-III). Technical report, Department of Science and Technology, UT Lakshadweep
4. Gadre MR (1989) Anti sea erosion works at Lakshadweep Islands site inspection report. Central Water Power Research Station, Pune

Chapter 4
Numerical Modelling of Coastal Processes of Kavaratti Island

Abstract A comprehensive study of wave climate and coastal processes during different seasons is required to delineate the factors responsible for the shoreline changes and to identify the locations that need protection. Numerical model is used as an effective tool for studying the nearshore wave climate and coastal processes. The model studies have been carried out using MIKE21 modelling system developed by the Danish Hydraulic Institute (DHI). The salient aspects in the model set up and the results of the simulations including some of the recommendations to minimise the coastal impact on Kavaratti in the Lakshadweep group of islands are presented in this chapter. The results of the modelling studies provide vital information for efficient formulation of disaster mitigation and management measures for protecting the islands.

Keywords Coastal erosion · Accretion · Numerical modelling · Waves · Current · Diffraction · Shore protection · Coastal processes

4.1 Introduction

Coastal erosion studies initiated by CESS as part of the beach monitoring programme [1, 2] have indicated significant shoreline changes in almost all the inhabited islands. The spatial variation in the erosion rates which are available from earlier studies as well as recent field observations are linked to a number of factors which include both natural and anthropogenic activities. The natural factors that contribute to the shoreline variations can be attributed to the effect of near shore waves, currents, winds, etc. However the annual net effect of these environmental forces on the shore remains more or less similar causing erosion during the monsoon season and accretion during the ensuing fair season, so that the equilibrium is well maintained. An exception to this natural condition is the occurrence of episodic events like severe depression/cyclones/storms etc., where the short period variation in shoreline particularly could be drastic such that total recovery of the

© The Author(s) 2015   61
T.N. Prakash et al., *Geomorphology and Physical Oceanography of the Lakshadweep Coral Islands in the Indian Ocean*, SpringerBriefs in Earth Sciences,
DOI 10.1007/978-3-319-12367-7_4

beach during the next fair season would not be possible even after a few years. The 2004 cyclone which affected Kavaratti is one such event. It was reported that the northern parts of the Kavaratti Island were badly affected during the passage of this cyclone. The high waves which lashed the island along with the coastal flooding due to heavy rainfall which lasted for 3 days had removed a major portion of the active beach on the northern region of the island, particularly the area facing the entrance channel [3]. The loss of material was so severe that the area continued to be an eroding coast even during the period 2007–2009 during which the study was carried out. In addition to this anthropogenic activities like removal of coral reef from the northern part as part of the development activities related to harbour development and construction of jetties on the western side of the island also have contributed significantly to the high rates of erosion. The erosion witnessed on the southwest coast of the island, adjoining the semi-enclosed lagoon can be directly linked to the limited supply of alongshore sediment from north to south during the fair season, due to the presence of two long jetties that act as groins preventing further movement towards the south. The gradual subsidence and attrition of the coral reef barriers that bound the lagoon due to high wave activity during the monsoon also undoubtedly have paved the way for increased wave activity on the western part of the island.

In this context a comprehensive study of the nearshore wave climate and coastal processes at work during different seasons is essential to delineate the causes responsible for the shoreline changes and also to identify the locations that need immediate protection. The present study aims at understanding these changes through numerical modelling. The model studies have been carried out using the MIKE21 modelling system developed by DHI. The salient aspects in the model set up and the results of the simulations are discussed in this chapter.

4.2 Numerical Model Studies: Models Used

Numerical model studies were carried out using various modules of MIKE21 modelling system. A brief description of the models/modules of MIKE21 suite of programs used for the present work is given below.

4.2.1 Spectral Wave (SW) Model

For the simulation of the wave climate the Spectral Wave (SW) model of MIKE21 was employed. MIKE21 SW is a new generation spectral wind-wave model based on unstructured meshes, which takes into account all the important phenomena like wave growth by influence of wind; non-linear wave-wave interaction; dissipations due to white-capping, bottom friction and depth-induced breaking; refraction and shoaling due to depth variations, and wave-current interaction [4].

The model simulates the growth, decay and transformation of wind-generated waves and swells in offshore and coastal areas. For the present study, a fully spectral formulation based on the wave action conservation equation [5, 6] was used. The basic inputs required for the numerical model are the bathymetry and the site-specific hydrodynamic data which include data pertaining to offshore waves, winds and bottom (sea bed) sediment characteristics. For preparing the bathymetric data, the nearshore fine grid data obtained from the Lakshadweep Harbour Engineering Department for the northern and western region was combined with the data from the charts prepared by National Hydrographic Office (NHO). For defining the offshore wave boundary condition the data from the offshore Wave Rider Buoy located approximately 25 km NW of Kavaratti Island (deployed by National Institute of Technology (NIOT), Chennai) was used. The outputs from the SW model are the wave parameters like the significant wave height, zero crossing wave period, mean wave direction, etc. The radiation stresses induced as a result of the wave action are also generated, as this data is required for further computation of currents and sediment transport using the Mike Flow Model FM.

4.3 MIKE21 Flow Model (MIKE21-FM)

To have a thorough understanding of the coastal processes and circulation of the study area, the hydrodynamic model for the model domain was set up using the Hydrodynamic Module (HD) of the MIKE21-FM modeling system. The Hydrodynamic Module has been developed for complex applications with oceanographic, coastal and estuarine environments. This model can be used to simulate a wide range of hydraulic and related items, including tidal exchange and currents, storm surges, heat and recirculation, water quality, coastal hydraulics and oceanography. It is based on an unstructured mesh with linear triangular elements and uses a cell-centered finite volume solution technique. The advantage of using an unstructured mesh is that it provides an optimal degree of flexibility in the representation of complex geometries and enables smooth representation of boundaries. This further helps in optimization of information required for a given amount of computational time. Hence providing a suitable mesh representing the model domain is the prime requisite for running the model.

Setting up of the model involves selection of the area to be modeled, generating the mesh model using the available bathymetric data so as to get the desired resolution, giving the flow, wind and wave fields considered for the study and also defining the boundaries. The basic input required for running the Flow Model are the domain and time parameters; initial and boundary conditions; calibration factors and other driving forces. The domain parameters are the bathymetry and computational mesh whereas the time parameters are simulation length and overall time step. The calibration factors for fine tuning of the model are bed resistance, momentum dispersion coefficients, wind friction factors, etc. For defining the initial conditions, if any, the water surface level/velocity components are given as input. Boundary conditions are defined by giving the water level/discharge along the model boundaries. The other driving forces that can be defined in the model are wind speed, wind direction, tide, source/sink of discharges and wave radiation stresses.

The outputs from the HD module are available in two forms—basic and additional variables for each element and time step. The basic variables are water depth and surface elevation; flux densities in main directions; velocities in main directions, temperatures, densities and salinities whereas the additional variables include current speed and direction, wind velocities, air pressure, drag co-efficient, precipitation/evaporation, courant/CFL number and eddy viscosity. The additional variables are generated only on request whereas the basic variables are available by default while running the model.

4.3.1 Sediment Transport Module

The sediment transport and related processes within the study area during the three seasons have been studied using the Sediment Transport (ST) Module available under the comprehensive MIKE21 Flow Modelling system mentioned earlier in the coastal circulation section. This module is used for the calculation of sediment transport capacity and related initial rates of bed level changes for non-cohesive sediment due to currents or combined waves and currents. The module calculates sediment transport rates on a flexible mesh (unstructured grid) covering the area of interest on the basis of the input data provided which includes hydrodynamic data (obtained from a simulation with the Hydrodynamic Module), wave data (provided by MIKE21 SW) and characteristics of the bed material. The simulation is performed on the basis of the hydrodynamic conditions that correspond to a given bathymetry. The ST Module covers the range from pure currents to combined waves and currents and also accounts for the effect of wave breaking.

The necessary input data for running the ST module are the model domain (bathymetry data and simulation length), hydrodynamic data (water depth and flow fields—from the Flow Model), wave data (wave height, period and direction from the SW Model), sediment properties (size and gradation of bed material) and morphology parameters (update frequency). Sediment transport rates and resulting morphological changes are the output data from the ST module. These are available in the form of basic output variables such as total load, bed level change, rate of bed level change and bed level.

4.4 Model Setup

Separate numerical models were set up for the three distant seasons viz. pre-monsoon (February–May), monsoon (June–September) and post-monsoon (October–January). This was required as the model calibration and validation needs to be done for each season as the tuning parameter used for calibration are not constant throughout and are subjected to seasonal variation.

4.5 Model Calibration and Validation

The hydrodynamic data pertaining to the pre-monsoon, monsoon and post-monsoon seasons collected by the field team are used for the calibration and validation of the numerical models. The recorded site specific hydrodynamic data after processing are critically examined to get first-hand information on the hydrodynamic characteristics of the island as it is primary data. Model calibration was carried out by comparing the model results with the field data collected during different seasons and adjusting the tuning parameters. The bed friction factor was the main parameter used for tuning.

The data from the studies on short-term beach changes in the Kavaratti Island conducted during December 2007–December 2008 with reference to the 20 beach monitoring stations viz. CSK 1 to 20 established by CESS in 2000 were also used for the validation of the modelling results. The results show very good corroboration with the field data.

4.6 Results and Discussion

4.6.1 Simulated Wave Parameters

The simulated maximum nearshore wave heights in Kavaratti Island, during the pre-monsoon, monsoon and post-monsoon seasons are shown in Figs. 4.1, 4.2 and 4.3. The average values of the simulated wave parameters for the three different seasons are also presented in figures. The simulated mean value of significant wave height during the pre-monsoon month of March is in the range 0.1–1.2 m. The maximum value of 1.2 m significant wave height occurs in the northeastern part i.e. in the coastal area in front of stations 6 and 7. The eastern part particularly the northeastern side of the island experiences higher wave activity compared to that of the lagoon coast. The mean significant wave height varies from 0.5 to 1.2 m in this region. There is a significant variation in the nearshore mean wave direction during the pre-monsoon period. The mean wave direction for the eastern part of the coast ranges from 140° to 180° whereas for the western coastal region bounded by coral reefs it is between 240° and 280°. The mean wave period during this period ranges from 5 to 9 s. For the monsoon month of July, the mean significant wave height is in the range 0.2–1.8 m. The northern and northeastern parts of the island are subjected to higher wave activity. Simulation results also indicate that the wave activity is less intense on the western coast even though the direction of approach of waves is 250°–270° during the monsoon. This is mainly due to the

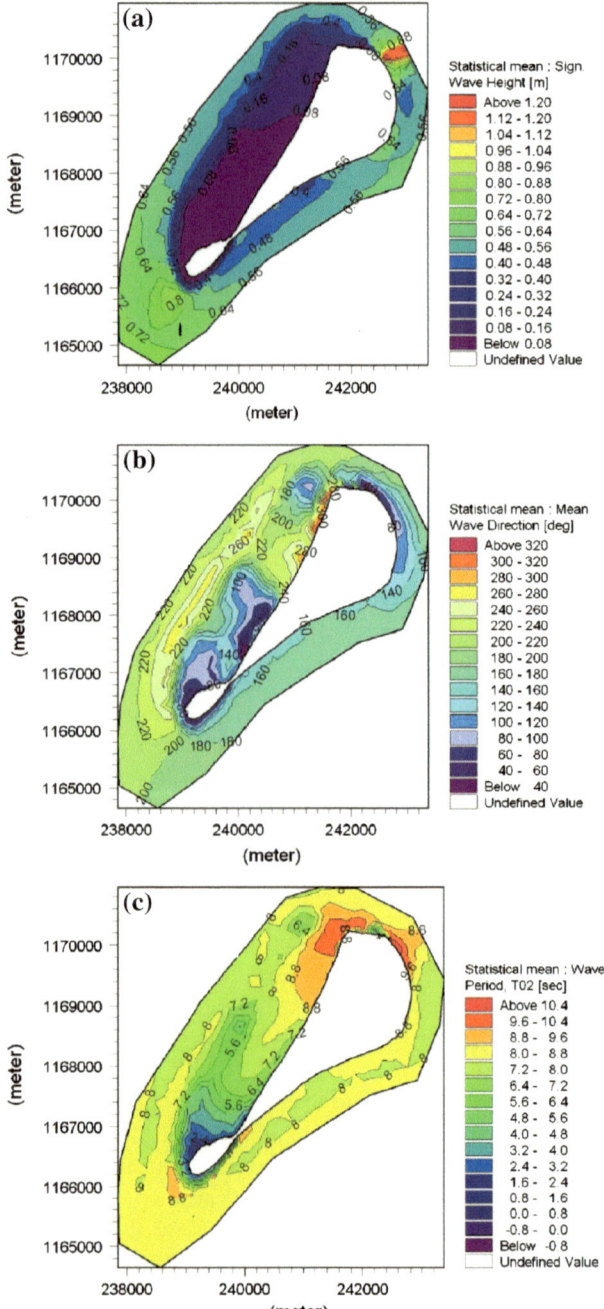

Fig. 4.1 Simulated mean values of wave parameters. **a** Significant wave height. **b** Mean wave direction. **c** Zero crossing period during the pre-monsoon month of March

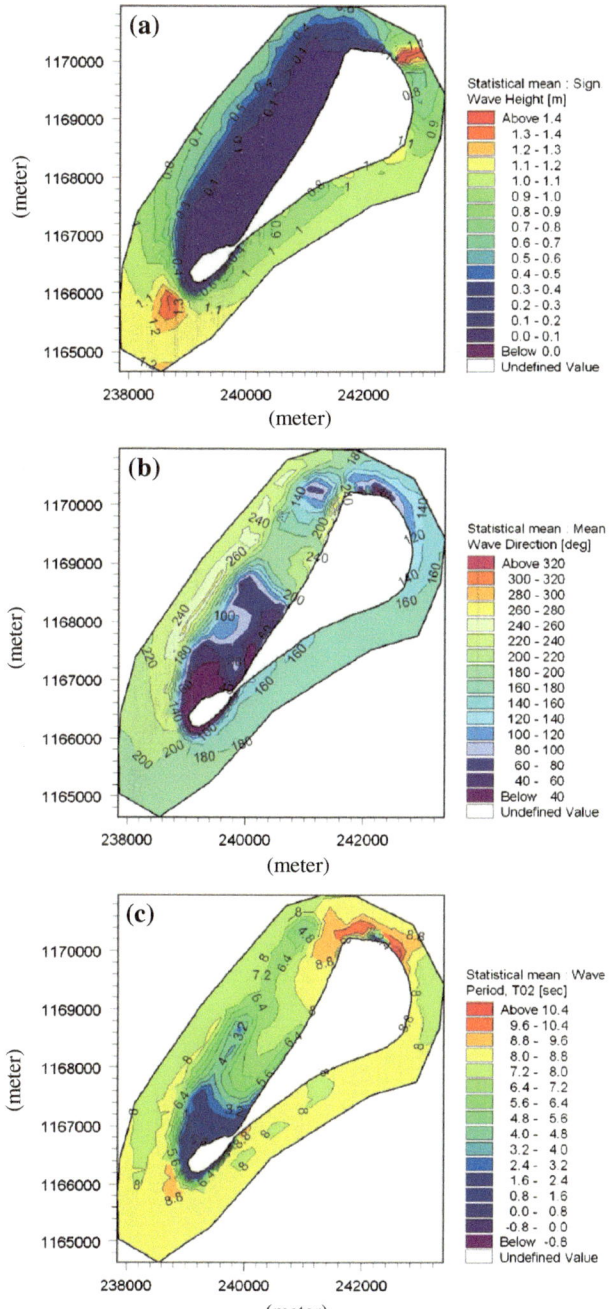

Fig. 4.2 Mean values of wave parameters. **a** Significant wave height. **b** Mean wave direction. **c** Zero crossing wave period during the monsoon month of July

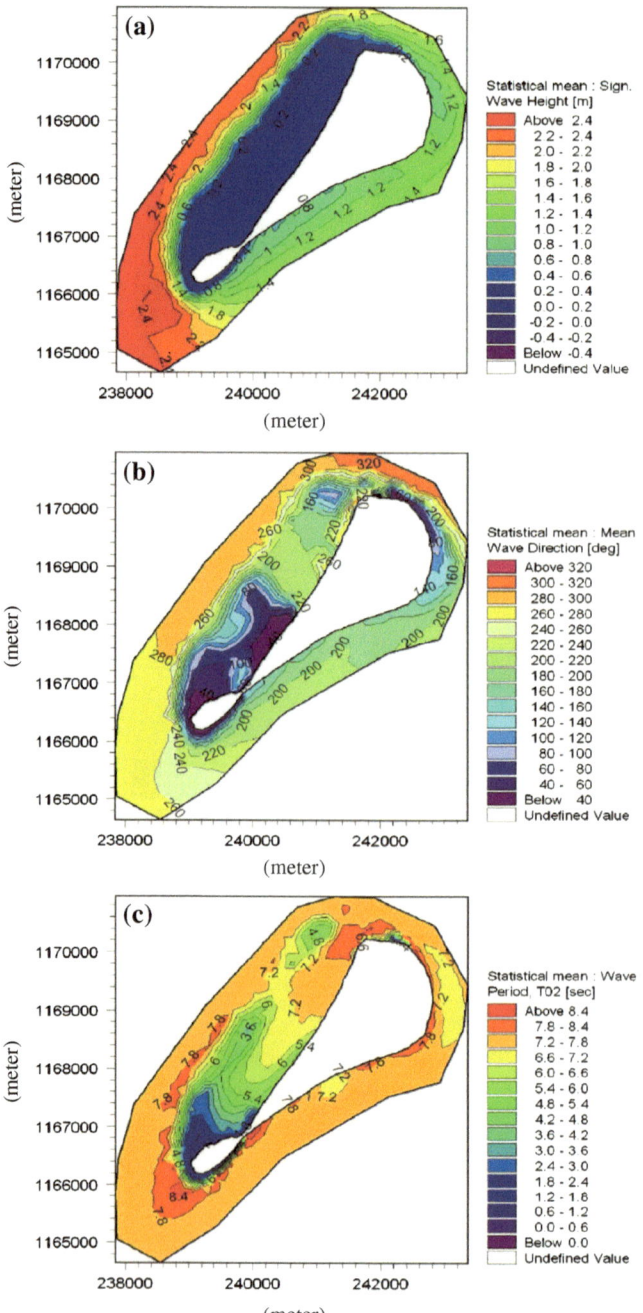

Fig. 4.3 Mean wave parameters. **a** Significant wave height. **b** Mean wave direction. **c** Zero crossing period during the post-monsoon month of November

natural protection given by the coral reefs which acts like submerged breakwater thereby attenuating the wave energy. As in the case of pre-monsoon there is a significant variation in the nearshore mean wave direction during the monsoon period. The mean wave direction for the eastern part of the coast ranges from 180° to 200° whereas for the western coastal region bounded by coral reefs it is between 220° and 280°. The mean wave period during this period is 7–8 s. For the post-monsoon month of November the mean significant wave height is in the range of 0.1–1.4 m. As in the case of the pre-monsoon and monsoon seasons, during this period also the northern and northeastern parts are subjected to higher wave activity. The eastern part particularly the northeastern part of the island experiences higher wave activity compared to that of the western lagoon coast. The mean wave direction for the eastern parts of the coast ranges from 160° to 180° whereas for the western coastal region bounded by coral reefs it is between 100° and 240°. The mean wave period during this period is 6–10 s. The statistical mean of the maximum wave periods for the various seasons also has been computed as this gives information regarding the swell waves. The statistical mean of maximum wave periods for all the seasons is in the range 12–15 s and for each season there is a distinct spatial variation of the of long period waves approaching the island.

4.6.2 Wave Diffraction

The effect of diffraction of waves on the Kavaratti Island, with its orientation in the SSW-NNE direction with a maximum length of 4.5 km (SSW-NNE direction) and width of 1.5 km (east-west direction) cannot be neglected. Hence separate studies were carried out to understand, analyse and study the effect of wave diffraction and refraction on the magnitude and direction of the nearshore waves, due to the transformation of offshore waves as they approach the island. The predominant wave directions and periods during the three distinct seasons were considered for the diffraction studies. During these seasons the prominent wave directions are 260°–270° and 200°–210° and dominant wave periods are 7–8 s [7]. Hence the analyses were carried out for two different cases—one with 265° wave direction and 7.5 s wave period and the second one with 210° wave direction and 7.5 s wave period. The results presented in Figs. 4.4 and 4.5 indicate that the southern part and the western part of the island are subjected to higher wave activity for both the cases. The wave activity due to diffraction effect on the eastern and northeastern part is comparatively less for both the cases. This observation is more or less similar to what has been reported earlier [7].

4.6.3 Coastal Circulation

The simulated mean current pattern during the three seasons—pre-monsoon, monsoon and post-monsoon are presented in Fig. 4.6. The analysis results indicate

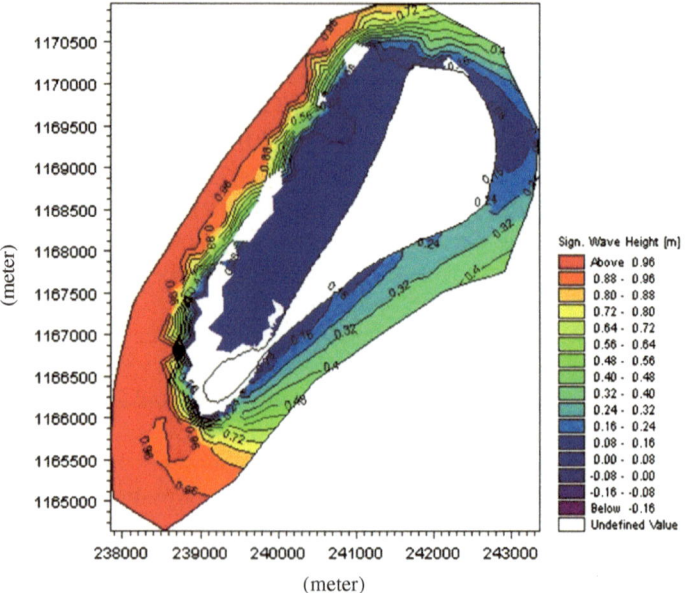

Fig. 4.4 Diffraction effect due to a waves of 1 m sig. wave height, 7.5 s wave period approaching from 265°N

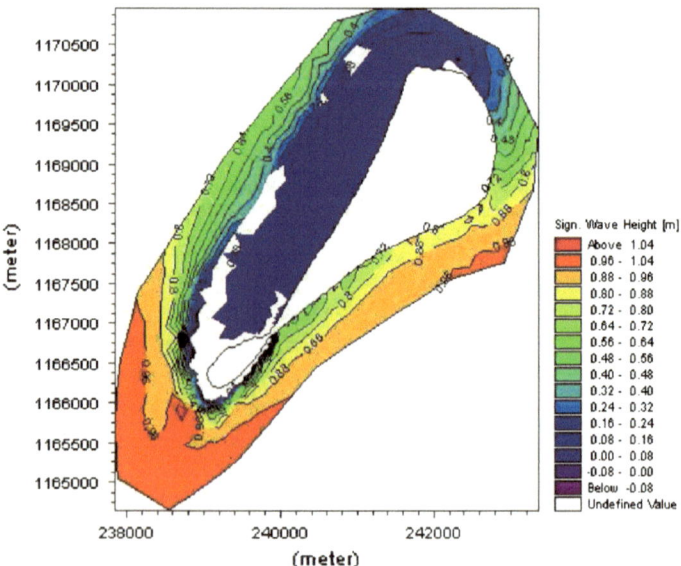

Fig. 4.5 Diffraction effect due to a wave of 1 m sig. wave height, 7.5 s wave period and approaching from 205°N

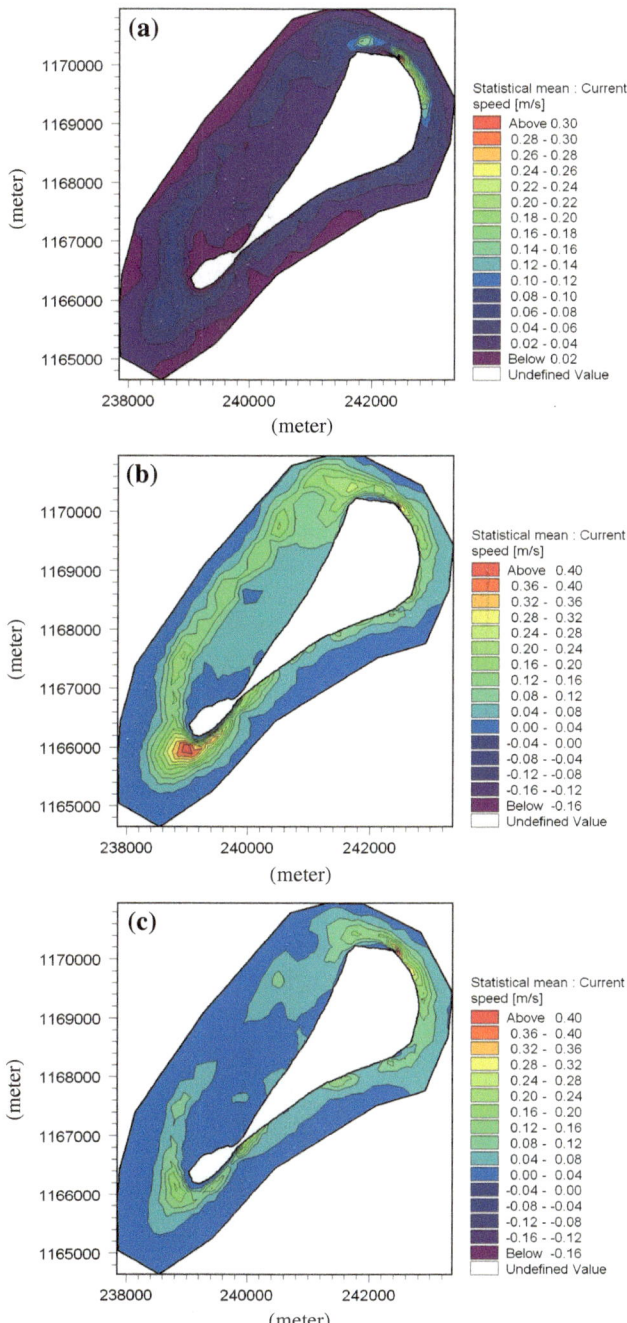

Fig. 4.6 Mean current speed during the **a** pre-monsoon month of March **b** monsoon month of July and **c** post-monsoon month of November

that the Northern part of the island especially the region extending from CSK 5 to CSK 8 experiences relatively higher currents during all the three seasons. The current in the North Western region (i.e. the coastal region adjoining CSK 9 and CSK 10) is also high during all the three seasons with the monsoon season giving the maximum impact. From Fig. 4.6b it is quite evident that during the monsoon the entire coastal stretch between coastal stations CSK 9 and CSK 11 is subjected to high currents. The proximity to the entrance channel and tidal influence can be considered as the main reason for the high currents observed in this region. Another interesting observation is that during the monsoon even though the pre-dominant wave direction is in the range 260°–270°, high currents are also seen on the eastern side of the island, especially the North Eastern and South Eastern region. This can be attributed to the refraction and diffraction effect of waves as they approach the shore during monsoon. However during the post-monsoon period, the refraction and diffraction effect is not appreciable on the western part of the island. This is because the western part of the island is protected to a large extent by the coral reef surrounding the lagoon area except for a few locations like the coastal region opposite to the natural inlet and near to the main entrance channel. At these locations the waves directly attack the coast. The maximum value of the mean coastal current from the simulation results is around 0.3 m/s in the northern region. Mean current direction inside the lagoon area varies between 80° and 140°; 80° and 100°; 100° and 120° during the pre-monsoon, monsoon, post-monsoon seasons respectively. The seasonal variation in the current direction is more pronounced on the eastern part when compared to the lagoonal area. For the east coast, the current direction varies between 200° and 240° during the pre-monsoon and post-monsoon seasons whereas the variation is between 60° and 120° for the monsoon season.

Typical circulation pattern in and around the Kavaratti island during the three seasons namely—pre-monsoon, monsoon and post-monsoon seasons are presented in Figs. 4.6 and 4.7 respectively. The longshore current direction inside the lagoon is predominantly towards north during the monsoon month of July whereas it is towards south during the post-monsoon month of November. Close examination of currents on the northeastern part and also along the eastern coast indicates oscillatory currents at one or two locations. This was also confirmed by the CESS project team involved in the longshore current measurement and other data collection. The observations of the field team are presented in Fig. 4.8. The current convergence tendency at one or two locations on the eastern side as indicated by the simulation results can be easily validated with the field team's report confirming the presence of one or two small pocket beaches.

4.6.4 Sediment Transport

The sediment transport rates for the three seasons were computed by activating the Sediment Transport Module while running the Flow Model of MIKE21. The total sediment transport rates for the Kavaratti coast during the pre-monsoon,

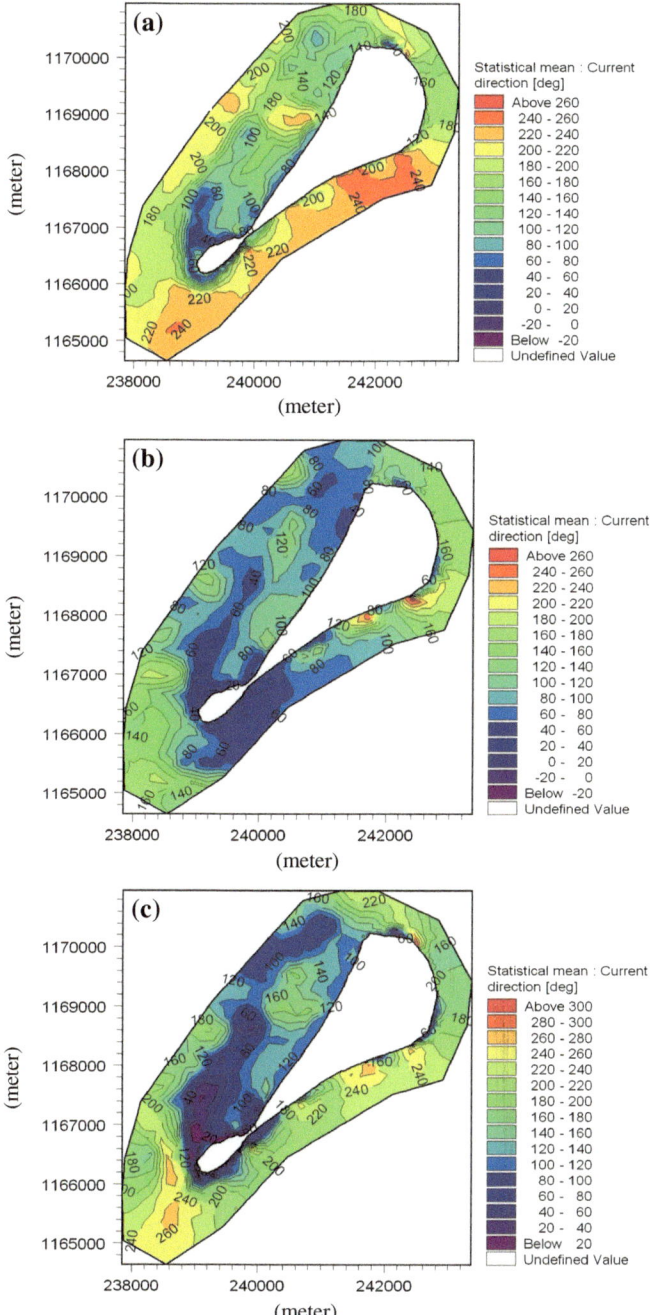

Fig. 4.7 Mean current direction during the **a** pre-monsoon month of March **b** monsoon month of July and **c** post-monsoon

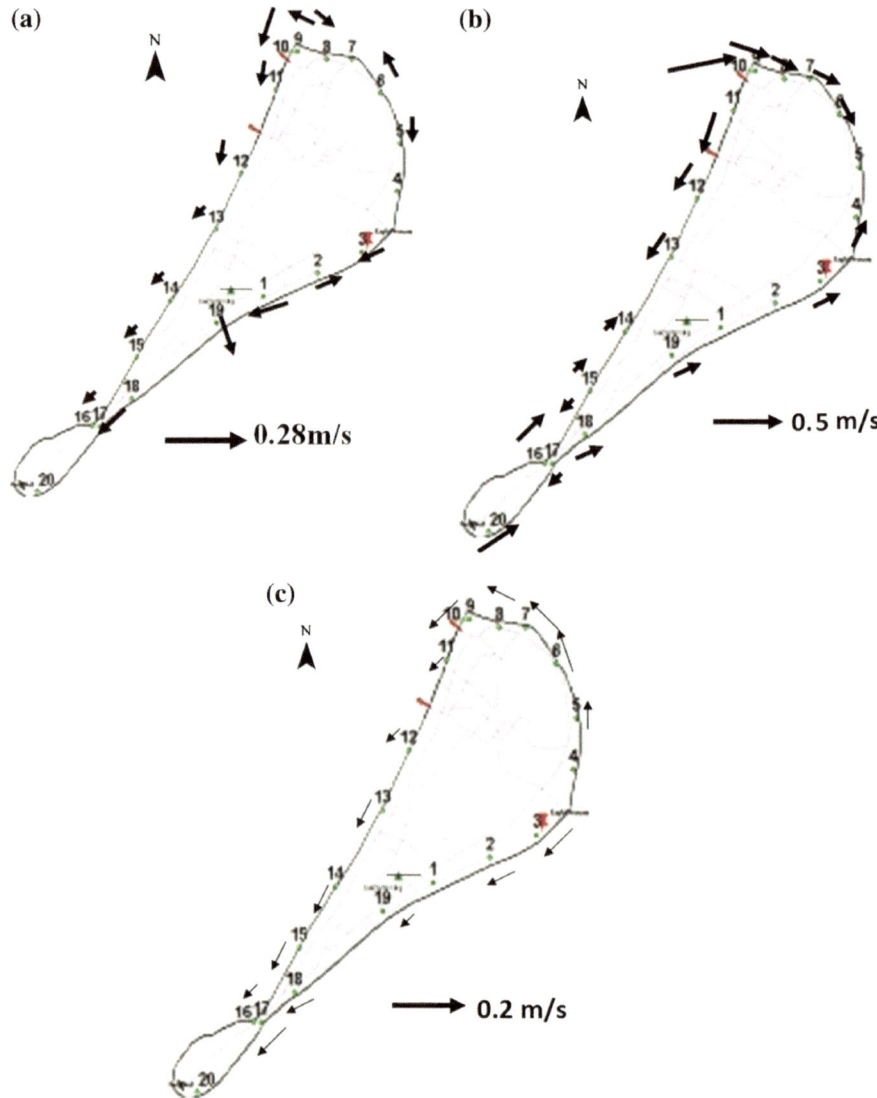

Fig. 4.8 Littoral observations made during the **a** pre-monsoon month of March **b** monsoon month of July and **c** post-monsoon month of November

monsoon and post-monsoon periods have been computed and are presented in Fig. 4.9 and the corresponding mean sediment transport directions are shown in Fig. 4.10.

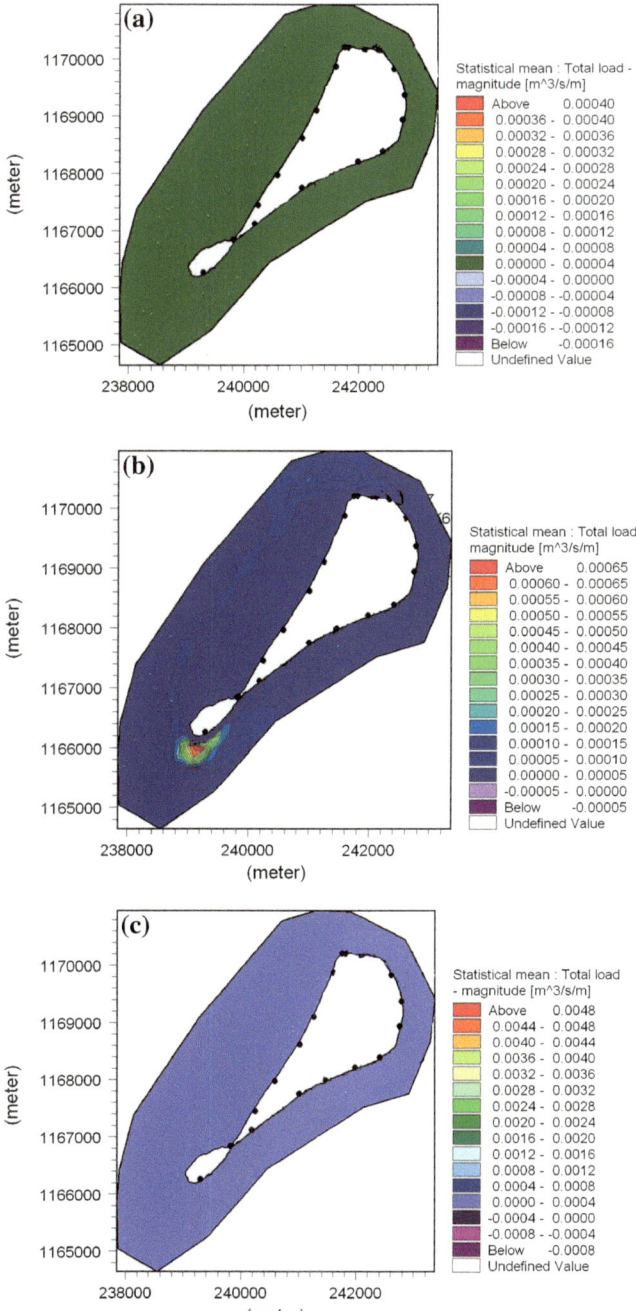

Fig. 4.9 Mean sediment transport magnitude during the **a** pre-monsoon month of March **b** monsoon month of July and **c** post-monsoon month of November

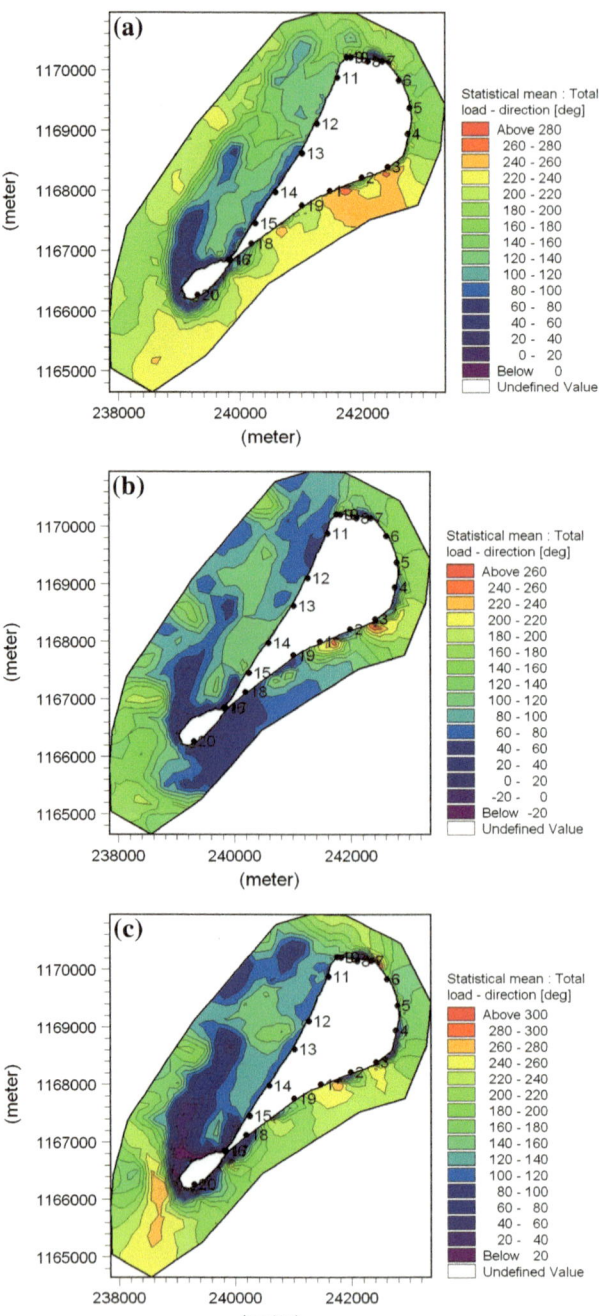

Fig. 4.10 Mean sediment transport direction: **a** pre-monsoon, **b** monsoon and **c** post-monsoon month of November

4.6.5 Bed Level Changes

These results are available along with the regular outputs of the sediment transport model. The mean value of bed level changes during the three seasons are presented in Figs. 4.11 and 4.12. On comparison it can be seen that the east coast of
the island is subjected to negative bed level change compared to that of the west

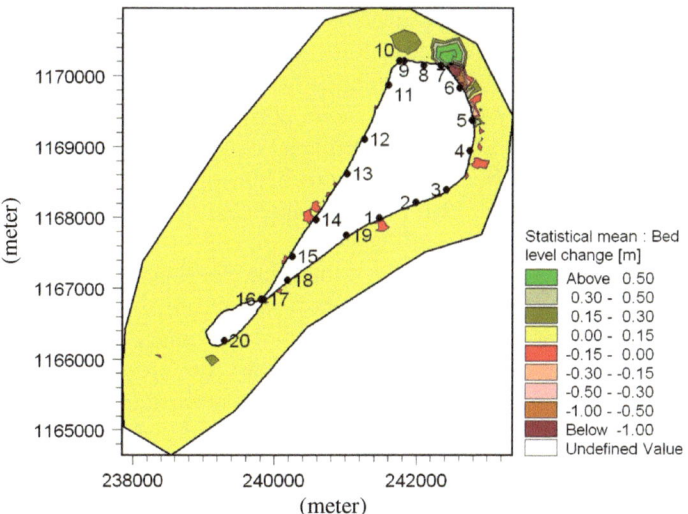

Fig. 4.11 Mean bed level change during the pre-monsoon month of March

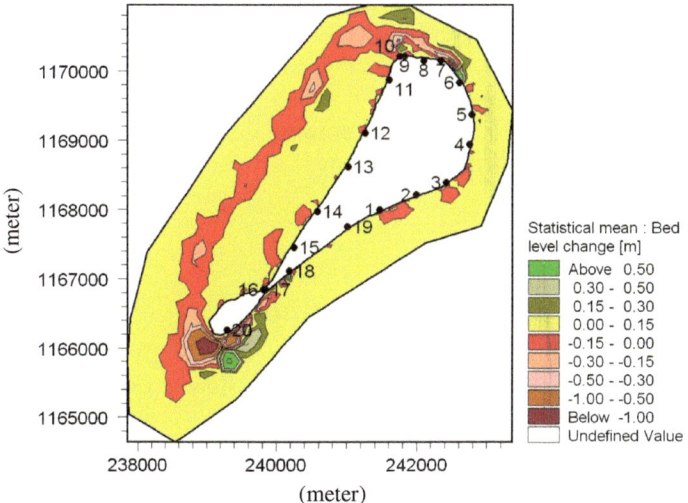

Fig. 4.12 Mean bed level change during the monsoon month of July

coast indicating higher erosion in this region. The beach in front of the coastal station CSK 8 is invariably an accreting beach even though there is a slight reduction in the accretion during pre-monsoon and post-monsoon. The maximum accretion is seen during the post-monsoon season and minimum during the monsoon. The accretion tendency is clearly due to the impact of the recent spit like growth that is seen adjacent to the station CSK 8. This growth essentially acts like a groin thereby protecting the adjoining coast towards the west. However the presence of this growth has a negative impact on the north-eastern part as it hinders the passage of sediment to the other side. Due to this there is a shortage of sediment supply to the eastern side and this is the main reason for the high erosion reported on the coast adjoining stations CSK 5–7. The simulation results also corroborate well with the field observations made by the Project Team presented in Tables 4.1 and 4.2. The validation of the simulation results for bed level change has been done by comparing with the data obtained from various sources like—plots showing short-term and long-term shoreline changes, field photographs taken during the different seasons (Figs. 4.13, 4.14 and 4.15).

On critically examining the simulated mean bed level changes for the three seasons (Table 4.3), it can be seen that the beaches to the north of CSK 5, in front of CSK 6 and the south of CSK 7 are eroding without much recovery even during the fair season, hence needs immediate attention. Site specific protective measures considering the terrain condition and nearshore wave characteristics have to be adopted. Of the remaining area, the beaches linked to CSK 9, 10, 17 and 20 can be categorized as areas under mild to moderate erosion. Critical examination of the bed level change analysis clearly indicates that the erosion in most of the cases is linked to the high monsoonal wave activity which is for a short duration of time. In most of the cases the negative impact due to this high wave activity is compensated by a more or less equal amount of deposition during the next fair season. However due to the fact that the temporal variation in wave pattern need not be the same every year there could be a very small net erosion or accretion in these locations. Appropriate, site-specific shore protection measures—hard or soft may be adopted for these areas. There is a marginal erosion observed near coastal stations, CSK 11, 12, 14, 15, 16 and the north of CSK 18. The eroding tendency of the beach on the lagoon side is primarily due to the scarcity of sediment coming to this region.

4.6.6 Impact of Anthropogenic Activities

The presence of the two jetties (which are operational) and one old jetty (in a dilapidated state) on the northern side of these regions has definitely had an impact on the sediment transport to the southern region. This is also quite evident from the field observations (beach profiles and photographs) made by the CESS project team.

Table 4.1 Shoreline characteristics during different seasons

Station	Beach orientation (°)			Foreshore slope (°)			Beach status
	1	2	3	1	2	3	
CSK 1	160	160	160	Steep	5	5.5	Eroding, protected by granite blocks and two layers of tetrapods
CSK 2	165	169	168	7	5	4.5	Eroding, protected by tetrapods, accreting during post-monsoon
CSK 3	145	148	148	Steep	4	3	Eroding, protected by tetrapods
CSK 4	108	110	110	6	4	4.5	Erosion prone, protected by tetrapods
CSK 5	90	98	–	8	5	5	Wide beach protected by tetrapods
CSK 6	60	60	65	Steep	–	4	Eroding during February and accreting during August
CSK 7	12	15	12	Steep	4	3	Highly eroding
CSK 8	15	15	15	7	4	3	Protected by tetrapods, accreting during December
CSK 9	20	18	18	Steep	4	2	Eroding and protected by tetrapods
CSK 10	310	321	301	–	4	3	Lagoon side with fine sand. Eroding towards north and protected by granite blocks. Accreting towards south
CSK 11	291	292	293	4	5	6	Accreting beach on the lagoon side with fine sand
CSK 12	305	299	300	4	5	5	Accreting beach on the lagoon side with fine sand. Cusps seen
CSK 13	302	310	310	6	6	6	Eroding beach on the lagoon side with fine sand
CSK 14	310	308	308	4	Steep	6	Eroding, protected by granite and concrete blocks
CSK 15	300	310	308	6	5	5	Stable, protected by tetrapods
CSK 16	320	315	318	5	3	6	
CSK 17	135	139	140	4	5	5	Sandy beach protected by granite blocks. Medium sand with shell fragments
CSK 18	145	155	149	Steep	Steep	3	Sandy beach with pebbles protected by tetrapods and hollow bricks during February and no beach in August
CSK 19	142	152	148	7	3	4	Slight accretion during February, protected by granite blocks and tetrapods
CSK 20	158	–	155	Steep	Steep	Steep	Eroding beach with high wave activity throughout the year

Note 1 pre-monsoon season, *2* monsoon season, *3* post-monsoon season

Table 4.2 Littoral environmental observations during different seasons

Station	Longshore current, speed and direction			Wave height (m)			Wave direction (°)			Wave period (s)		
	1	2	3	1	2	3	1	2	3	1	2	3
CSK 1	OC	OC	OC	2	1	0.75	150	165	153	14	10	10
CSK 2	0.06 252	OC –	0.15 240	1.5	1	0.75	165	165	168	12	12	10.5
CSK 3	0.37 250	0.15 50	0.17 230	1.5	1	0.75	155	164	153	12	12	9.5
CSK 4	OC	0.16 15	OC	<1	1	1	88	111	113	12	11	9.5
CSK 5	OC	OC	OC	<1	1	1.75	85	85	98	7	12	8
CSK 6	OC	0.12 160	0.14 330	1.3	1.3	1.8	40	58	68	6	9	7
CSK 7	OC	0.15 118	0.22 305	1	0.65	1.65	10	25	68	6	9	8
CSK 8	0.22 280	0.09 120	0.19 290	<1	0.5	0.4	20	42	53	6	10	8
CSK 9	0.04 290	0.2 115	– N	–	0.4	0.25	10	27	48	8	9	–
CSK 10	0.04 220	0.37 75	0.21 223	–	–	–	350	350	–	–	–	–
CSK 11	0.11 210	0.12 210	0.07 225	–	–	–	–	295	–	–	–	–
CSK 12	0.11 220	0.13 220	0.07 210	–	–	–	–	300	–	–	–	–
CSK 13												
CSK 14	0.12 215	0.07 45	0.15 210	–	–	–	–	310	–	–	–	–
CSK 15	0.09 220	0.08 40	0.12 210	–	–	–	–	315	–	–	–	–
CSK 16	OC	0.03 40	0.08 220	–	–	–	–	310	–	–	–	–
CSK 17	0.1 52	0.05 240	0.17 225	<1	0.75	0.8	–	–	–	–	–	–
CSK 18	0.09 59	0.1 60	0.11 225	1.5	1	0.75	153	140	148	14	11	10
CSK 19	0.14 245	0.11 70	0.11 213	3	1	0.5	165	142	148	15	12	10
CSK 20	0.11 63	0.39 N	OC	<1	1	0.75	195	194	153	11	10	10

Note 1 pre-monsoon season, *2* monsoon season, *3* post-monsoon season, *OC* oscillatory current, *N* northerly

Fig. 4.13 Photographs showing the beach conditions at the monitoring stations during pre-monsoon

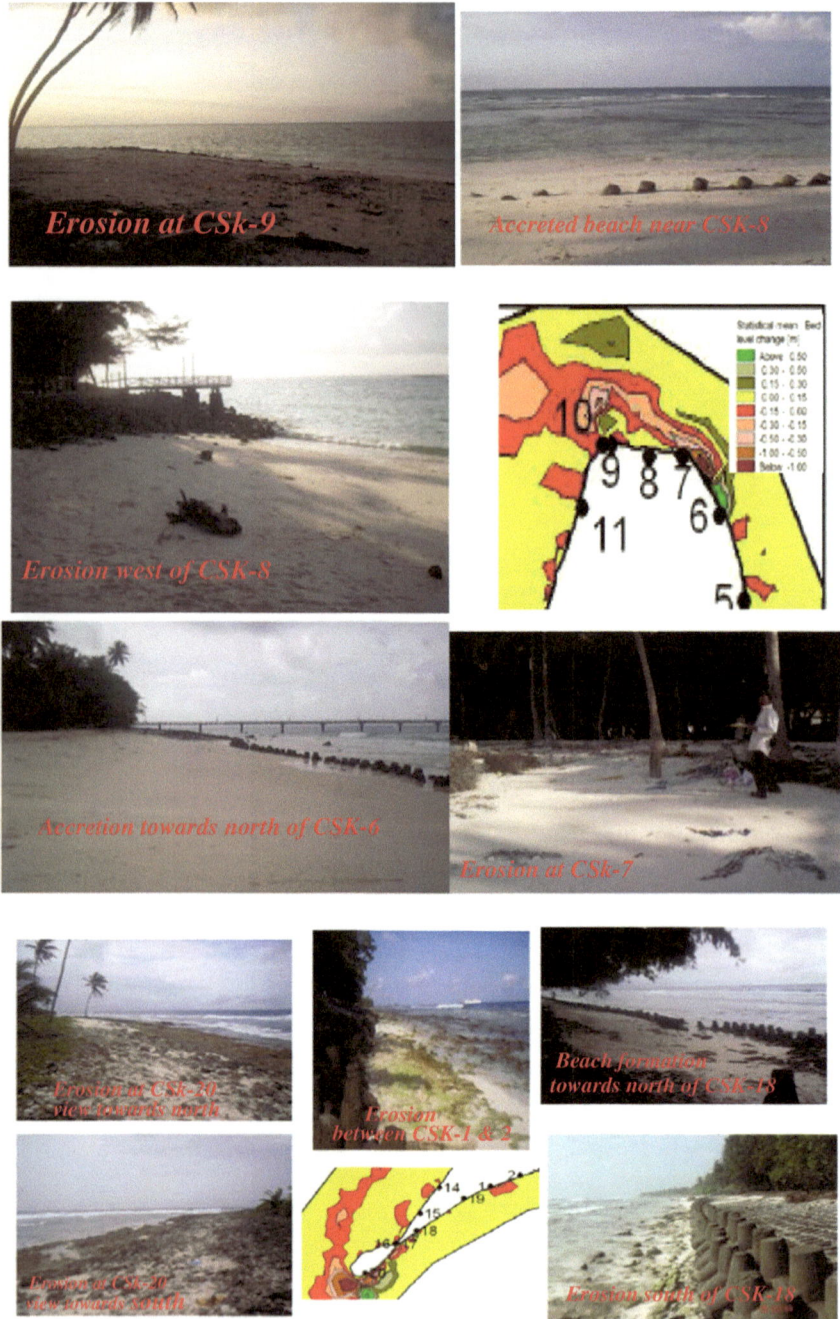

Fig. 4.14 Photographs showing beach conditions at the monitoring stations during monsoon

Fig. 4.15 Photographs showing beach conditions at the monitoring stations during post-monsoon

Table 4.3 Beach condition at various coastal stations based on numerical modelling results

Coastal station	Beach status			Remarks
	Pre-monsoon	Monsoon	Post-monsoon	
CSK 1	Mild erosion	Mild erosion	Mild erosion	Needs protection
CSK 2	Stable	Stable	Stable	Needs no protection
CSK 3	Stable	Mild erosion	Mild erosion	Needs protection
CSK 4	Stable	Stable	Stable	Needs no protection
CSK 5	High erosion towards north	Mild erosion towards north	Moderate erosion towards north	Highly Eroding, needs protection
CSK 6	High erosion	Accretion towards north	High erosion	Moderate erosion, needs protection
CSK 7	High erosion towards south	High erosion	High erosion	Highly Eroding, needs protection
CSK 8	Accretion	Accretion	Accretion	Protection not required
CSK 9	Stable	Mild erosion	Moderate erosion	Moderately eroding, needs protection
CSK 10	Stable	Mild erosion	Mild erosion	Soft measures can be adopted for combating seasonal erosion
CSK 11	Stable	Erosion	Stable	Dynamically stable, protection not required
CSK 12	Accretion	Erosion	Erosion	Dynamically stable, protection not required
CSK 13	Accretion	Erosion	Accretion	Eroding, needs protection
CSK 14	Mild erosion	Mild erosion	Mild erosion	Soft measures can be adopted to combat mild erosion
CSK 15	Mild erosion towards south	Accretion	Mild erosion towards south	Soft measures can be adopted to combat mild erosion in the south
CSK 16	Stable	Mild erosion	Mild erosion	Seasonal erosion can be prevented by adopting soft measures
CSK 17	Stable	Moderate erosion towards south		Critically eroding, needs immediate protection
CSK 18	Moderate accretion	Mild erosion towards north	Moderate accretion	Seasonal erosion, soft protection measures sufficient
CSK 19	Stable	Stable	Stable	Needs no protection
CSK 20	Stable	Erosion	Erosion	Needs protection

4.6.7 Shore Protection Measures

The results of the numerical model studies clearly indicate that the erosion observed at many locations is linked to the high monsoonal wave activity which is for a short duration of time. In most of the cases the negative impact due to this high wave activity is compensated by a more or less equal amount of deposition during the next fair weather season. However due to the fact that the temporal variation in wave pattern need not be the same every year there could be a very small net erosion or accretion at these locations. Soft shore protection measures are required in these areas. Marginal erosion is observed near coastal stations—CSK 11, 12, 14, 15, 16 and the north of CSK 18. The eroding tendency of the beach on the lagoon coast is primarily due to the scarcity of sediment coming to this region. Remedial measure for this area would be to consider the option of sand by-passing to the southern side of the jetties or placing of geotubes. The geotubes can be placed either along the shore or underwater (submerged) very near to the shoreline to prevent further receding of the shoreline. Since these areas are not so critical, cost effective protection measures which give optimum performance can be considered.

4.6.8 Strengthening of Reef on the Northern Part of the Island

In addition to the above work, based on the request from DST, UTL separate numerical model studies were conducted to study the feasibility of strengthening the existing reef on the northern region. Both short-term and long-term shoreline changes were modelled using the LITPACK model of DHI. On the basis of the model results it can be concluded that the reef strengthening would stabilize the beach in front of coastal stations CSK 8 and CSK 9. However there are indications that the shoreline to the east of station CSK 8, i.e. near station CSK 7 is likely to experience severe erosion if the proposed shore protection measure is connected to the shore. This problem can be easily solved by ensuring that sediment by-passing is allowed on either side of the proposed structure. There are also indications of seasonal erosions at beach stations CSK 9 and CSK 10. Careful analysis of the results, points out that the main contributing factors are direct exposure to the main entrance channel located very near to this and scarcity of sediment reaching the shore which can be attributed to the blocking of sediment passage created by the jetties on the lagoon coast. Adopting suitable remedial measures to allow free movement of trapped sediments could be an environment friendly soft measure that would definitely improve the condition. Another option worth considering is to artificially nourish the area with the dredge spoil from the main entrance channel which is being regularly dredged for maintaining the required depth for navigation.

4.7 Summary

Numerical modelling for conducting a comprehensive study of the wave climate and coastal processes at work in the Kavaratti Island has been carried out using the MIKE21 software developed by the DHI. The salient aspects in the model set up and the results of the simulations are presented in this chapter. The results of the numerical model studies clearly indicate that the erosion observed at many locations is linked to the high wave conditions and also due to anthropogenic activities. Based on the results of the study the critical areas have been identified and recommendations for the protection of these areas are also addressed.

References

1. Prakash TN, Suchindan GK (1994) Coastal erosion studies in the Lakshadweep Islands. In: Proceedings of Indian National conference on harbour and ocean engineering, Pune, pp D31–D41
2. Prakash TN, Shahul Hameed TS, Suchindan GK (2001) Shoreline dynamics of selected islands of Lakshadweep archipelago in relation to wave diffraction. Geol Surv Ind Spec Publ 6:201–209
3. Prakash TN, Shahul Hameed TS (2004) Site inspection report (Cyclone 2004) in Lakshadweep Islands. Department of Science and Technology, UT Lakshadweep
4. DHI Manual (2004) Danish hydraulic Institute user manual and reference guide for MIKE21 and LITPACK modules
5. Komen GJ, Cavaleri L, Donelan M, Hasselmann K, Hasselmann S, Janssen PAEM (1994) Dynamics and modelling of ocean waves. Cambridge University Press, Cambridge
6. Young IR (1999) Wind generated ocean waves. In: Battacharyya R, McCormick ME (eds) Ocean engineering book series. Elsevier, Amsterdam
7. Baba M, Shahul Hameed TS, Kurian NP, Subhashchandran KS (1992) Wave power of Lakshadweep Islands. Final report submitted to Department of Ocean Development, Government of India, Centre for Earth Science Studies, Trivandrum

Chapter 5
Energy Resources

Abstract The electrification of the Lakshadweep Islands was started during the Second Five Year Plan period (1956–1961) and all the inhabited islands were electrified by the end of the Sixth Five Year Plan (1980–1985). The main source of power generation is diesel which is transported from the main land in large quantities and is stored in barrels. The cost of power generation is very high compared to main land. The non-conventional energy sources like solar, bio-mass and wind power can be alternative resources in the islands. Due to its geographical position, the solar energy is available throughout the year except during the rainy monsoon season. During this season the wave and wind energy is highest which could be an alternative. Wave power potential of Lakshadweep is 40 % more than that of the main land. Preliminary studies on the economics of wave power indicate that the cost of wave power generation becomes comparable with the existing rates for which the fuel has to be transported from the mainland. A multi-source power generation system is proposed for the Lakshadweep Islands.

Keywords Non-conventional energy · Wave and wind power potential · Multi-source · Power generation

5.1 Introduction

The electrification of the Lakshadweep Islands was first taken up during the Second Five Year Plan period (1956–1961). Minicoy was the first island in the Lakshadweep group to be electrified in the year 1962 followed by Kavaratti in 1964 [1]. Subsequently all the remaining inhabited islands were electrified by the end of the Sixth Five Year Plan (1980–1985). The main source of power generation has been diesel generators in almost all the islands which are equipped with stand-alone power generating devices without any inter island connection in the supply grid. During the period 1962–1982 the power supply which was initially limited to 6 h daily in all the islands, except Kavaratti where 24 h power supply

© The Author(s) 2015   87
T.N. Prakash et al., *Geomorphology and Physical Oceanography of the Lakshadweep Coral Islands in the Indian Ocean*, SpringerBriefs in Earth Sciences,
DOI 10.1007/978-3-319-12367-7_5

Table 5.1 Island wise maximum demand (kW) during the last five years (2007–2012)

No.	Island	2007–2008	2008–2009	2009–2010	2010–2011	2011–2012
1	Minicoy	955	1,037	1,107	1,175	1,200
2	Kavaratti	1,320	1,452	1,548	1,649	1,750
3	Amini	665	710	748	788	850
4	Andrott	925	994	1,082	1,144	1,100
5	Kalpeni	498	538	563	591	650
6	Agatti	650	695	740	788	850
7	Kadamat	598	632	670	710	950
8	Kiltan	329	360	375	390	500
9	Chetlat	255	265	280	295	400
10	Bitra	42	44	50	54	60
11	Bangaram (inhabited)	45	45	70	72	20
	Total demand (kW)	6,282	6,772	7,233	7,656	8,330

was available since 1964. It was made available round the clock with 100 % electrification of the island in 1983. All the activities related to the power generation, transmission and distribution of electricity in the Lakshadweep Islands are managed by the Lakshadweep Electricity Department which has its headquarters at Kavaratti. The maximum demand for power is around 8.33 MW and it includes the requirements of both domestic and commercial consumers. Table 5.1 gives the maximum demand for the power in the 11 islands (10 inhabited and a tourist centre) during the period 2007–2012.

5.2 Present Status of Power Generation

The present power needs of the Lakshadweep Islands are met from the diesel and solar power generators (Figs. 5.1 and 5.2). The installed capacity of diesel generators are 9.97 MW and that of solar power generator is 0.76 MW. Other sources of power like the wind, bio-mass, etc. also have been explored. Island wise installed capacity of diesel generators is presented in Table 5.2. The small size of the islands, remote location and the non-availability of resources locally for the generation of power by conventional methods are some of the factors that have affected the pace of infrastructure development activities. In this context it is imperative that the power generation has to be expanded in such a way that the islands become self reliant which will definitely pave the way for better economic growth. The islands being located far away from the nearest main land (Cochin/Calicut/Mangalore depending on the location of the island) the present cost of power generation is extremely high compared to the other states of India. This is mainly because of the additional charges involved for transportation of fuel, lubricants and other items from the mainland to the respective islands for

Fig. 5.1 View of the diesel power station at Kavaratti Island

Fig. 5.2 View of a grid interactive solar power plant

power generation. Normally excess quantities of oil are stored in each of these islands to ensure uninterrupted power supply, especially during adverse weather conditions where transport by sea is not feasible. Availability of adequate space for storage of fuel which is usually brought in barrels of 200 L capacity is another major constraint for further expansion of the existing diesel plants. The risk involved in handling the hazardous fuel also is very high as the entire fragile eco-system consisting of corals and other rare species of flora and fauna will be badly affected in the event of occurrence of any mishap like fire or oil spill.

Table 5.2 Island wise installed capacity of the diesel generator (DG) sets

S. No.	Name of the island	Installed capacity (kW)
1	Minicoy	2,800
2[1]	Kavaratti	3,200
3	Amini	1,900
4	Andrott	3,250
5	Kalpeni	1,250
6	Agatti	2,350
7	Kadmat	2,400
8	Kiltan	1,000
9	Chetlat	500
10	Bitra	120
11	Bangaram (uninhabited)	120
Total		18,890

The relatively high cost of power generation through diesel plants and its potential risk on the fragile ecosystem of the islands have prompted the Lakshadweep authorities to consider other alternative sources of power like the non-conventional and renewable sources of energy. Of the various non-conventional sources solar, bio-mass, and wind power were tried initially on experimental basis. Based on the performance of the pilot plants, steps were taken to upgrade the system. Of the various alternative methods considered, the solar power generation gave reasonably good results and was successfully adopted in some of the small islands to meet the daily power requirements to a certain extent. Table 5.3 gives the installed capacity of solar power in the Lakshadweep. Till 2000 the installed solar plants were functioning as standalone. Subsequently many of the old plants

Table 5.3 Installed capacity of solar photovoltaics

S. No.	Island	Date of commissioning	Capacity (kWp)
1	Kiltan	02/03/2000	100
2	Minicoy	27/11/2000	100
3	Agatti	02/02/2002	100
4	Kadmat	08/03/2002	150
5	Andrott	31/03/2002	100
6	Kavaratti	26/04/2002	100
7	Bitra	20/01/2004	50
8	Kalpeni	02/05/2004	100
9	Suheli	02/05/2005	15
10	Bangaram	09/06/2005	50
11	Chetlat	02/03/2009	100
12	Amini	09/05/2009	100
Total installed capacity (Figures are as on 1.9.2009)			106

were augmented to higher capacities and integrated with the existing Diesel Grid System to cater to the additional requirement so as to ensure an uninterrupted power supply. Bigger plants require larger space for installation and getting the required space is a major constraint in the Lakshadweep due to acute scarcity of land. Also the non-availability of continuous solar power of the required capacity during the monsoon period which lasts for 2–3 months (during June–August) is another drawback to consider it as a sustainable perennial source. Hence the other renewable clean energy sources like wind and waves have to be considered.

5.3   Economics of Power Generation

The estimated power requirement for the Lakshadweep group of islands for the year 2013–2014 is around 50 MU [1]. Studies on power consumption in the islands indicate that there is an increase in demand of nearly 9–10 % every year. The present cost of power generation is around Rs. 30/kWh, out of which around Rs. 20/kWh accounts for the cost of fuel and lubricants. This is certainly very high compared to the production cost of Rs. 4/kWh through conventional means in the main land. Instead of taxing the public the government is providing power at a highly subsidised rate of Rs. 4/kWh which undoubtedly is a big burden for the Lakshadweep Electricity Department.

The transportation cost of the diesel/lubricants from the mainland contributes to a sizeable percentage of the total cost. The diesel required for power generation is transported from Calicut in Kerala to the respective islands and stored in barrels. The official estimate shows that about 66 lakh litres of diesel per annum is being purchased. Because of the relatively small land area and also the high density of population of each island, the storage of oil which is a hazardous material in large quantities is highly risky as it may lead to major disasters like fire, contamination of ground water, soil, etc. in the event of oil spill. The transportation of oils in large quantities in the barges also poses a serious threat to marine flora and fauna, particularly the corals in the surrounding waters. The corals are the most biologically rich and economically valuable ecosystems of the Lakshadweep which have already started depleting due to various reasons. Noise and air pollution due to the continuous operation of the Diesel Generator sets is another important issue which needs attention. Considering all these negative aspects which certainly will affect the fragile environment of all the islands and also the present cost of power generation, there is an urgent need for looking into technically and economically viable alternative sources of clean renewable energy that will have less impact on the environment. In this context it would be worth considering the renewable sources of energy from waves, solar, wind, bio-mass technology, etc. Out of these solar powers generation is being attempted/adopted in a very small scale in all the islands. But the available space for installation of solar panels is a major limitation mainly because of the small size of islands and also the presence of thousands of tall coconut trees which is one of the major sources of income for the islanders. Attempts have also been made to tap other sources of renewable energy like wind and bio-mass. But the performance of the wind mills was not that satisfactory as the winds are not strong enough to generate significant quantity of power except during the monsoon and occurrence of episodic events like storms. The presence of tall coconut trees in all the islands is another issue of great concern as the height of the land based wind masts should be at least 25 m to capture the wind energy without any interference. Increasing the design height of the wind mast is not always technically viable as there is a limitation on the maximum depth of the foundation that can be adopted for a structure in the island for safely transferring the load to a hard stratum for fixity. In this context ocean wave power generation seems to be a better option mainly because it is a clean, environment friendly technology and it is available throughout the year. Other advantages in the adoption of this clean energy technology are the non-requirement of storage space, and absence of risks in handling hazardous oil, less environmental impacts on the fragile eco system like hazard of fire, air, noise and water pollution.

Coming to the economics part of it the per unit cost of generation is comparable with the current rates as the present power generation is mostly dependent on the diesel generated power for which the entire requirement of diesel and lubricants are brought from the mainland.

5.4 Wave Power Potential

Wave power can be defined as the transport of energy by ocean surface waves. Specifically, it is the energy transmitted per unit length of the wave crest in a direction perpendicular to the wave approach. The wave power (P) in kilowatts per metre length of the wave crest can be computed using the relation [2].

$$P(kW/m) = 0.55\,H_s^2 Tz$$

where H_s is the significant wave height in metres and T_z is the zero crossing period in seconds. Studies carried out by Centre for Earth Science Studies (as discussed in Chap. 2) had revealed that the Lakshadweep Sea has the potential for wave power generation [3]. The monthly distributions of wave power in the Lakshadweep Sea are depicted in Fig. 5.3. During the monsoon months there is an increase in wave power in the Lakshadweep Sea and the period from June to August stands out with higher wave power occurrences. During June half of the distribution is in the range 20–35 kW/m and in July it is in the range 30–45 kW/m. By August three-fourth of the wave power is spread over the range 10–35 kW/m. During November–March the most frequently occurring wave power is in the lowest range 0–5 kW/m and during April, May, September and October it is in the next range 5–10 kW/m. January and February are the calmest period with almost all the wave power confined to the range 0–5 kW/m.

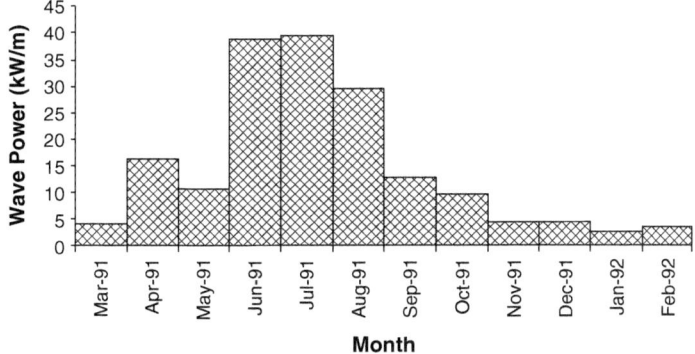

Fig. 5.3 Monthly distributions of the wave power (*Source* CESS Buoy data)

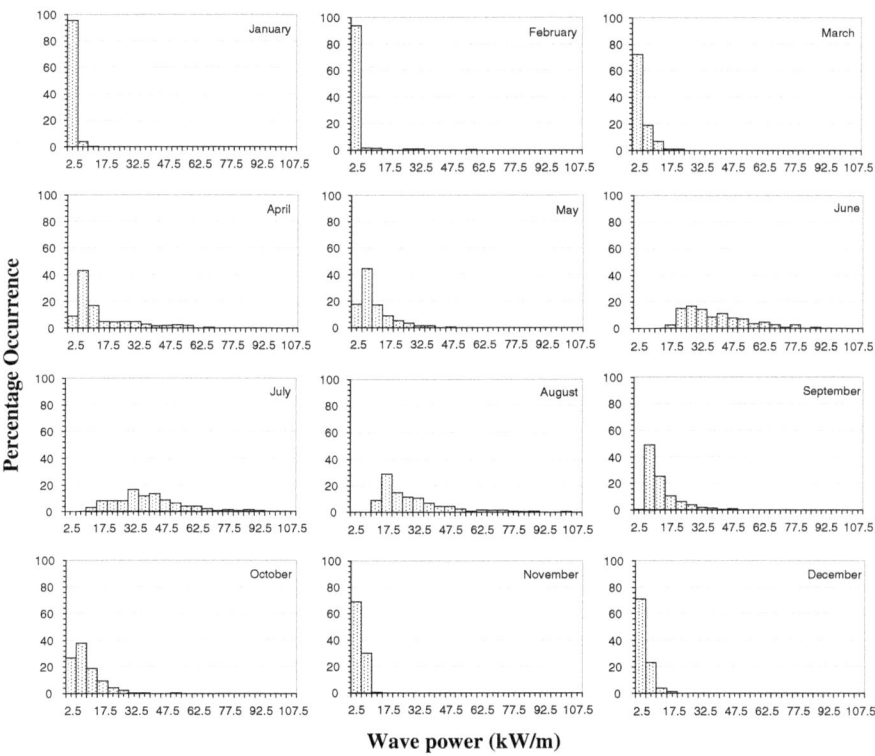

Fig. 5.4 Monthly average of wave power in the Lakshadweep Sea (*Source* CESS Buoy data)

The monthly average wave power is presented in Fig. 5.4. The mean wave power ranged from 28 to 40 kW/m during the months of June–August. For other months, they are in the range 2.6–16.2 kW/m. The monthly average wave power is more than 10 kW/m during the period April–September. The lowest mean wave power (<5 kW/m) is observed during November–February. The yearly average wave power for the Lakshadweep Sea is 13.9 kW/m which is about 27 % more than the yearly average for the location of maximum wave power potential along the coasts of the main land [3–6]. The maximum average wave power for the location along the main land is 25 kW/m during June and July, whereas it is nearly 40 % more for the Lakshadweep Sea (34.9 kW/m) and it persists during the period June–August. Offshore wave characteristics recorded by the NIOT Wave Rider Buoy (DS2) deployed at a depth of 1,800 m, approximately 25 km northwest off Kavaratti Island, also show similar trend (NIOT).

5.5 Wind Power Potential

Another untapped source of renewable energy which is available in the Lakshadweep is the wind energy. Since the islands are tiny in nature, the wind blowing is characteristic in speed to the sea shore and is almost the same as for all the islands, as in the case of solar radiation. The monthly average wind speed at Agatti Island is presented in Table 5.4 together with the solar radiation. The wind speed is higher during monsoon months and is in the range 8.4–8.9 m/s. The speeds are less during the remaining period, falling in the range 3.1–4.3 m/s except in September when the average speed is 5.9 m/s.

The monthly mean solar radiation and wind speed are given in Table 5.4 for a comparison. It can be seen that the trends of solar radiation and wind speed are opposite in nature. When the solar radiation is increasing the wind speed is found decreasing and vice versa. The maximum wind speed is during the months of June and July, whereas the minimum solar radiation is observed during the same period. Except in January this reversing trend can be observed throughout the year. This

Table 5.4 Monthly average solar radiation and wind speed at Agatti

Month	Solar radiation (kWh/m^2/day)	Mean wind speed (m/s)
January	5.127	3.4
February	5.765	3.7
March	6.270	3.2
April	6.043	3.9
May	4.984	4.3
June	3.926	8.9
July	3.779	8.9
August	4.268	8.4
September	4.946	5.9
October	4.682	3.7
November	4.727	3.4
December	4.672	3.1
Average	4.932	5.1

shows that there is good scope for these two sources of energies to be coupled for electric generation (solar-wind hybrid generator) to attain a more or less steady power along with the other power sources.

5.6 Advantages of Renewable Energy

The prime advantage of adopting the renewable energy is that it is clean and the risk involved is comparatively less as it can be located in the offshore. The wave and wind power plants can be installed conveniently at locations very close to the shore as all the inhabited islands have a lesser sloping region present in the nearshore [3]. For cost effectiveness the installation of these plants can be combined (as shown in Fig. 5.5) or with other major projects like establishment of desalination plants, jetties and harbours, etc. As the distance from the shore can be considerably reduced the cost of installation of the structure like platform or mooring for the floating device can be brought down considerably and also the transmission cost. At present three of the inhabited islands viz. Kavaratti, Agatti and Minicoy have desalination plants of 1 lakh litre capacity installed on the eastern side of the island (Fig. 5.6a–c). All these plants have an intake sump located at 7 m water depth and 24 h power supply is required for the continuous operation of these plants. At present these plants are solely dependent on Diesel Generators for power which invariably contributes to the high cost of production of potable water. There can be significant reduction in the cost of power, if power from wave/wind/solar energy plants is also utilized at these locations. Other options like hybrid renewable energy plants which combine wave energy, solar power and wind energy with by products like desalinated water would be highly beneficial.

Fig. 5.5 A multi-source (wind-solar-wave) power generator system. *Source* http://www.finavera. com/

Fig. 5.6 View of the desalination plant installed by NIOT in **a** Kavaratti, **b** Agatti and **c** Minicoy

5.7 Summary

The highest wave power is observed during the months of June–August, the range of values being 11–110 kW/m. During other months the power ranged up to 70 kW/m. The peaks of the distributions of wave power are in the ranges 20–25 kW/m during June, 30–35 kW/m during July and 15–20 kW/m during August. The wave power potential of the Lakshadweep Sea is 27 % more than that of the place with maximum wave power potential in the main land. In terms of wave power availability, the Lakshadweep Sea has more potential for wave power generation compared to the coasts along the main land.

Wave power which is a renewable source of clean energy can be considered as a technically and economically viable alternative source of energy for the Lakshadweep group of islands. Preliminary studies on the economics of wave power indicates that the cost of wave power generation becomes comparable with the existing rates as the present power production is mostly dependent on diesel generated power for which the fuel has to be transported from the mainland. Even though other alternate sources of the renewable energy like wind and solar power are also being adopted there are certain limitations in these methods mainly due to the reduced solar power during the peak monsoon period and the reduced power from wind mills during the relatively calm periods with low wind speed of less than 2.8 m/s. In this context even stand alone wave power plants can be economically viable for the remote islands of the Lakshadweep as it is the only perennial source locally available. The steep nearshore slope available on the eastern side of islands is an added advantage as the wave power plants can be located at moderately deep waters at a short distance from the shore and the transmission cost can be considerably reduced.

References

1. Basic Statistics (2007) Draft eleventh five year plan 2007–2012. Planning and Statistics Department, Secretariat, UT Lakshadweep
2. Shaw R (1992) Wave energy—A design challenge. Ellis Horwood Publishers, New York
3. Baba M, Shahul Hameed TS, Kurian NP, Subhaschandran KS (1992) Wave power of Lakshadweep Islands. Report submitted to Department of Ocean Development, Government of India, Centre for Earth Science Studies, Trivandrum
4. Baba M, Thomas KV, Shahul Hameed TS, Kurian NP, Rachel CE, Abraham S, Ramesh Kumar M (1987) Wave climate and power off Trivandrum. Project report on sea trial of a 150 kW wave energy device off Trivandrum coast. IIT, Madras on behalf of Department of Ocean Development, Government of India
5. Shahul Hameed TS, Kurian NP, Baba M (1994) Wave climate and power off Kavaratti, Lakshadweep. In: Proceedings of Indian national dock, harbour and ocean engineering conference, CWPRS, Pune, pp A63–A72
6. Baba M, Shahul Hameed TS, Kurian NP (2001) Wave climate and wave power potential of Lakshadweep Islands. Geol Surv India Spec Publ 56:211–219

Chapter 6
Integrated Coastal Zone Management Plan for Lakshadweep Islands

Abstract The Lakshadweep Islands are subjected to different ecological, economic and natural hazard vulnerabilities. Integrated Coastal Zone Management (ICZM) is an essential tool for tracking the different issues and managing the resources in a sustainable way. These plans become all the more important with implementation of the Coastal Regulation Zone (CRZ) Notification, 1991 by the Ministry of Environment and Forests (MoEF), Government of India. ICZMP were prepared for the Lakshadweep Islands following the issues and resource based approach through field mapping supported with high resolution satellite imageries. Integration and analyses of data were done using Geographic Information System. Identification of sites for different development activities in each island was proposed based on stakeholders' perception and EIA studies. Management options for the major issues confronting the islands are also discussed.

Keywords Integrated coastal zone management (ICZM) · Resources · Hazards · Coastal regulation zone · Sustainable use · Lakshadweep Islands

6.1 Introduction

Many countries around the world are working on the concept of Integrated Coastal Zone Management (ICZM) to manage the resources, multitude of uses, inherent hazards and to mitigate the natural hazards. The ICZM is defined as a continuous and dynamic process by which decisions are made for the sustainable use, development and protection of coastal and marine areas and resources by ensuring that the decisions of all sectors and all levels of government are harmonized and are consistent with the policies of the nations [1]. This concept has been worked out for many coastal stretches in the country [2] including the Lakshadweep group of islands [3].

The islands even though have long conjured up images of 'paradise', their amazing lagoons and coral reefs show signs of increasing stress. As the island communities strive to raise the living standards in tune with the increase in

© The Author(s) 2015   99
T.N. Prakash et al., *Geomorphology and Physical Oceanography of the Lakshadweep Coral Islands in the Indian Ocean*, SpringerBriefs in Earth Sciences, DOI 10.1007/978-3-319-12367-7_6

population, there is always a tendency to disturb the fragile ecosystems, which is one of the most valuable assets. At times there is a tendency to overexploit these natural resources and pollute the environment. Another aspect is the rising sea levels which can be attributed to the global climate change which are likely to damage the coastal areas and even submerge some of the low-lying islands. This will certainly affect the island economies with a negative impact on fisheries, tourism, coral reefs and freshwater resources. Islands are also important contributors to global biodiversity as the lagoons and coral reefs are home to many rare species. But there are indications that these environmentally sensitive habitats are under increased stress which may badly affects the flora and fauna and in case of some of the native endangered species it may even lead to an irreparable loss. The ICZM Plan for the islands is intended to preserve the functional integrity of the islands ecosystem, reduce the resource/or issue based conflicts arising out of the developmental activities in and around the island and to minimise damages to the coral ecosystem for the sustenance of the islands.

6.2 Coastal Zone Management in the Lakshadweep

Site specific scientific observations on coastal processes, shoreline changes and impacts of natural disasters are required to develop appropriate management plans and provide interventions if needed as part of the ICZMP. The National Centre for Earth Science Studies (formerly CESS), Trivandrum was one among the institutes which generated an exhaustive field data on nearshore waves, coastal morphology, sediment transport and shoreline changes of the Lakshadweep Islands to form the baseline data for ICZMP.

6.2.1 Existing Management Regulations

The only legislative instrument for controlling the coastal land use in the country is the Coastal Regulation Zone (CRZ) Notification issued in 1991 by the Government of India, under the Environment Protection Act [4]. Subsequently this was modified and replaced with the Coastal Regulation Zone [5] for the mainland and Island Protection Zone [6] notifications for the islands. The main objectives of the above Notification are to protect the people and property from coastal hazards and to prevent the degradation of the coastal ecosystems. Coastal Zone Management Plans (CZMP) have been prepared for Lakshadweep following the CRZ [7] notification. The Integrated Island Management Plan (IIMP) for the islands is being prepared following the guidelines given in [6] by using the concepts and principles of Coastal Regulation Zone [5].

6.2.2 Local Level Planning

Top-down planning and implementing mechanism often fail due to the lack of support and commitment within local communities. In contrast, the three-tier Panchayati Raj System (see also Chap. 1) of governance and planning which has already taken firm roots in the country actively encourages the local communities to participate as principal stakeholders. The elected Panchayat council of the island will play an active role in the implementation of the programmes/schemes suggested as part of the proposed ICZMP.

6.2.3 Coastal Zone Management Plan (CZMP)

As a continuum to the notification, the Union Territory of Lakshadweep prepared its Coastal Zone Management Plans (CZMP) demarcating the HTL, LTL and CRZ categories [8]. While preparing the CZMP, Centre for Earth Science Studies proposed a new method for identifying High Tide Line (HTL) based on geomorphic signatures in the field which made the demarcation of the HTL easy and simple. Even though the CZMP prepared for the islands had different CRZ categories (CRZ I to CRZ IV) which covered ecologically sensitive, developed, rural and water areas, it had the drawback of not addressing the various environmental and developmental issues. This has been addressed in the ICZMP that has been developed for the Lakshadweep coastal zones during the period 2003–2005.

6.3 ICZMP Approach

ICZMP preparation is based on resource and issue based approach. Information on the land use and assets are the basic inputs required for developing the management plan of a particular region. In case of Lakshadweep group of islands, the corals being the main ecosystem are the basic inputs and this is further supplemented with spatial and temporal data on the various physico-environmental characteristics of the islands. The preparation of ICZMP involves two steps. The preparation of conservation map that aimed at appropriate allocation of areas for the conservation and development is the first step. Preparation of environmental appraisal map with the status of environment with respect to major ecosystems and processes formed the second step. This formed the basic document to plan intervention and strategies to ensure quality environment. The conservation and environmental appraisal maps along with other physico-environmental characteristics of the island were synthesized in ARC-GIS. Sites earmarked for different activities in the island like settlement, development, tourism, fisheries, coastal protection, sewage disposal etc., were also identified based on stakeholder's perception and Environment Impact Assessment (EIA) studies carried out as part of the development programmes listed in the Xth plan document of the Lakshadweep.

6.4 Physico-environmental Characteristics of the Islands (Resource)

In order to prepare the ICZM plans the spatial and temporal data on the various physico-environmental features such as land use, landform, assets, beaches, inter-tidal flats, lagoon waters, built-up land etc. were mapped on 1:4,000 scale. The assets included government buildings, hospitals, schools, dwelling sites, fish landing centres, jetties, coastal protection structures, etc. Agriculture and fishery were considered as the major resources of the islands. Socio-economic data was also collected to bring out the financial status of the people in the islands. In addition to this the elevation of the island with respect to Mean Sea Level (MSL) was also generated to understand the micro-geomorphologic features of the island.

6.5 Issues/Problems Related to Islands

Majority of the islands are facing many issues/problems. They have been prioritized and the most important ones requiring immediate management interventions are listed below:

6.5.1 Coastal Erosion and Shore Protection

Coastal erosion is a serious problem faced by the islands every year. With a tiny land area of 32 km^2 and land elevation of 0.5–6 m above MSL, erosion of even a small portion of the land can be a significant loss. It has also been reported that the predicted sea level rise due to global climate change could be as high as one meter in the next century and can lead to flooding, land erosion, and or even submergence of some of the low-lying islands. Occurrence of storm/severe depression due to passage of cyclone can also cause extensive damage to the islands' coast. The unprecedented flooding and erosion of some of the islands in the Laccadive group during May 2004 due to the impact of a cyclonic storm [9] was devastating. These coral islands have an added advantage over the other low-lying tropical islands as the fringing reefs which grow at depths of 8–10 m can act as the first line of defence whenever there is a high wave attack. The islands are coral atolls which are generally protected from the open sea by live corals that fringe the islands and if sea level rise doesn't occur at an alarming high rate and if the coral continues to grow taller at the same rate without any mortality the island will be protected to a certain extent. During the 7th five-year plan period (1986–1990) Lakshadweep Islands witnessed a spurt of activities like construction of jetties, port and harbour development, widening of entrance channel, etc. As a result of these developmental activities the coastal zone of many of the islands were badly affected. The removal of reef edge which was considered inevitable for widening of entrance channel and other construction works related to harbour development activities had eventually affected the stability of the shoreline which was otherwise dynamically stable. Replacing the removed reef through artificial reef is one of the management options proposed to restore the stability of the coast. The long-term shoreline change for all the inhabited islands is given Chap. 3. The analysis of implementation schemes for the coastal protection is given in Table 6.1.

Continuous coastal monitoring programme is needed both before adoption and after implementation of the ICZM Plan as it provides vital information on how the coastal systems are changing due to both natural and anthropogenic activities. This information is utilised for taking the management decisions at various levels.

Table 6.1 Implementation schemes/management options for coastal erosion

Issues	Management option	Factors to be considered for feasibility
Many areas have been protected by tetrapods and low-cost shore protection structures	Continue with low-cost shore protection/maintenance work	Many adverse impacts—loss of beach, affects the tourism industry, shore based fishery, etc.
	Retreat or moving to safer location	Retreat is not possible since the land area is less. Partial relocation can be worked out
	Beach nourishment	Studies on the design beach width, dredged sand can be used, sand resource in the lagoon bed to be identified
Low cost shore protection structure caused loss of frontal beaches and damage to tourism/fishery industry	Groins	Careful design is needed as lee-side erosion is a side effect, helps in beach formation on one side or in between two groins
	Artificial reef (offshore)	Prioritize areas of reef construction; develop proper reef design, which can induce the formation of beach. It helps to regenerate beach and offers recreational facilities for tourists

6.5.2 Fresh Water Management

Availability of potable water is a serious issue that requires immediate management intervention because of the limited groundwater resources. Based on the
present status of groundwater availability and demand, the islands of Kavaratti,
Minicoy, Agatti and Amini are categorised as deficit whereas Chetlet and Kalpeni
are likely to become deficit in the near future. In view of the above scenario, good
water management practices like water recycling, rainwater harvesting, creation of
awareness among the people, etc. need to be adopted to sustain the scarce water
resources of the islands. The Administration has already started taking appropriate measures like promoting rainwater harvesting (making it mandatory), put-
ting limits on pumping of groundwater etc. Desalination of water using the Low
Temperature Thermal Desalination (LTTD) principle is one of the viable options
for the augmentation of water supply to the island.

6.5.3 Conservation of Coral Reefs/Mangroves

The lagoons and corals are two distinct elements of nature of which the lagoons
are ecologically important as they houses a wide range of species. They supply
the essential nutrients and provide a safe breeding ground for the young species.
The most likely sources of pollution to the lagoons are from the land-based activities. Dredging activities in connection with the port development also affects the
lagoon water circulation. Deepening of lagoons to create artificial navigation chan-
nels and, harbour must be controlled. In the management option a buffer zone
above the HTL has been suggested to control sewage and storm drainage effluents
from the island, to safeguard against runoff soils from the island, etc. If develop-
ments on the lagoon shore are not planned properly it creates a variety of short and
long-term economic losses.

Available scientific literature shows that up to 70 % of the world coral reefs
will be destroyed in the next 20–40 years if destruction continues at the present
rate. Current activities undertaken by different institutions on coral reefs research
are sporadic and isolated, mainly because of the lack of physical facilities. Another
approach that has proved effective for coral reefs surrounding small islands is the
development of a marine reserve and sanctuary. This model basically encourages
the local communities to be responsible for their fishery and coral reef resources.
The reserve model provides the limited protection for coral reef and fishery surrounding the entire island but safeguards from all extraction or damaging activities in the area where the coral reef coverage exceeds 20 % [10]. This reserve and
sanctuary approach can provide real benefits to local fishing communities in the
island through increased or stable fish yields from the coral reefs that are maintained and protected. The resource management must be routed through the local
island communities to conserve their own marine resources. In summary, effective

coastal resource management in the islands is more than a problem of environmental consideration or law enforcement. Community based approaches involving the people who directly depend on these resources can give better outputs as strict regulatory mechanism alone will not yield desired results. Combining community, environmental surveys and legal approaches in a manner appropriate for a particular island would give better results. The churning motion of the boat propellers in the shallow water is one of the causes for the resuspension of sediments. A fraction of the coral mortality in the island is ascribed to the smothering effect of the resuspended sediment. One of the recommendations is to explore the utilization of fan-driven crafts—of the type 'Swampbuggy'—for shallow water traffic.

6.5.4 Fishery Resource Exploitation and Catch Enhancement

Tuna fishing is the main commercial fishing activity in the island. During fair weather period, the fishermen even migrate to nearby uninhabited islands which are identified as sites with high tuna potential. More than 25 % of the present landing is accounted from these uninhabited islands. The traditional pole and line fishing method, which was indigenously developed some decades back, is still being practiced for Tuna fishing. There is hardly any improvement in the methods adopted even though very efficient techniques are available worldwide. Shortage of live bait is deeply felt among the fishermen and this may be the major constraint in the intensification of tuna fishing. Shortage of adequate cold storage facilities in the islands is another matter of concern and this has forced the fishermen to bring out value added products like Masmin (smoked and dried fish) using the excess fish. The value added products from Shark like processed products from shark fins, oil extraction from liver etc. have immense potential which remain untapped. Mariculture has not been attempted in a major way till now in spite of the vast lagoons offering a good scope of growing a variety of species including the pearl oyster. Though the Administration has already made a beginning on the culture of ornamental fishes still there is scope for further improvement.

Non-availability of proper storage facility and lack of demand in the domestic market are the main reasons for the under exploitation of fishery resources. Any developmental programme aimed at improving the economy and employment opportunities of the Lakshadweep has to be invariably linked to the fishing industry as it is the main source of revenue. In this context, conceptualization and implementation of viable projects targeting the unexploited fishery resources including both tuna and other resources deserve top priority. Simultaneously setting up of basic infrastructure for proper storage, preservation and transportation to commensurate with production is also essential for the long-term development of fisheries sector in the Lakshadweep Islands. The implementation schemes/management options pertaining to the fishery sector are listed in Table 6.2.

Table 6.2 Implementation schemes/management options for fishery sector

Issues	Management option	Factors to be considered for feasibility
Fishery as the major occupation	Continue as such	Will not bring any substantial improvement in economy
	Improve and provide better infrastructural facilities	Further improvement is possible and would enhance employment opportunities
Diversification	Mariculture and value added products from fish	Development of value added fish products including the development of pearl oyster and ornamental fishery

6.5.5 Sewage and Solid Waste Treatment

Sewage and solid waste disposal is a major problem faced by the islands. The increasing population pressure and related development activities have put enormous pressure on the island ecosystem. The island communities create considerable amounts of sewage waste (50,000–1.2 lakh litre/day), which is often left into septic tanks, or cesspools that leach organic matter and pathogenic bacteria into the relatively shallow fresh water lens. To meet the daily requirement of fresh water the inhabitants of the islands mostly depend on groundwater, which floats as a thin lens on seawater under equilibrium conditions. Often these ground water sources are exposed to bacterial contamination due to the porous nature of the soil, shallowness of aquifer and proximity to leach pit/septic tanks which is a major risk to human health.

The sewage which is rich in nutrients causes high rate of eutrophication in lagoon waters resulting in excessive weed growth and depletion of the dissolved oxygen content thereby affecting the normal growth of corals and other organisms. Further, the pathogenic bacteria often associated with sewage will render the coastal waters unfit for water sports and other recreational activities. With the drastic increase in sewage that is being generated it is expected that by 2025 the quantity will be doubled. In the absence of proper sewage disposal system there is every possibility that the fresh water lens would be contaminated. The Island Administration is aware of the danger and is making concerted developmental measures to mitigate the problem by adopting methods like introducing bio-toilets.

Another important environmental issue that needs to be addressed is the disposal of solid non-biodegradable waste. As such there is no organized waste disposal system in the islands. Due to this solid waste materials like plastic, polythene, glass materials, etc., accumulate on the islands and gradually spill over into the adjacent waters causing serious threat to the lagoon eco-system. At present a simple system for the collection and disposal of non-biodegradable waste is being adopted which is not that effective. Through this scheme the waste is collected and deposited at specially created garbage depository and shipped to mainland for re-processing/land filling etc.

For the disposal of bio-waste a system of compost processing can be implemented and this has already being carried out in phased manner. The most appropriate mode of disposal appears to convert the biodegradable waste into bio-fertilizer by open pit aerobic composting process.

6.5.6 Tourism

The Lakshadweep group of islands with beautiful white sandy beaches, crystal clear lagoon waters rich in corals and surrounded by aquamarine deep waters is an important tourist destination both for domestic and international tourists. It is one of the sources of income. But being categorised as an ecologically sensitive area with limited natural resources there is certain restrictions imposed by the administration for protecting and preserving its pristine natural eco-system. The visits of tourists are restricted by issuing permits for their period of stay prior to the visit. Ship services to these islands are available from the port of Cochin in Kerala. In addition to this speed boats ply between the major islands providing inter-island transport facility. During the monsoon season of June–September ship services are limited. Agatti is the only island which has an airstrip and daily flight services are there from the mainland (Cochin). At present only limited infrastructure facilities are available in the island. Even though there is scope for further expansion it is being done in a controlled way mainly to preserve the sensitive eco-system with its rich flora and fauna and also considering the availability of resources like land, water, electricity, food etc. The carrying capacity of each island is another important factor to be considered for deciding the maximum number of tourists that can be permitted.

Hence while preparing the ICZMP due consideration has been given for all these factors for arriving at an efficient tourism development and management plan (Table 6.3). Based on the ICZMP approach locations/areas suitable for tourism can be identified by providing additional infrastructure facilities if required.

Table 6.3 Analysis of implementation schemes for tourism

Issues	Management option	Feasibility
Tourism as a major source of revenue	Continue with better infrastructure facilities	Positive impact
	Opening the uninhabited islands for increased tourism	It is good for island economy, better employment opportunities, minimal environment impact and optimum resource consumption
Depletion of potable water and use of limited resources	Rain water harvesting and sewage treatment plants	Improves the potential
Increase the water related activities in the lagoon	Promote more water sports and scuba diving centres in the islands	Develop uninhabited islands which has the potential

The land area being considerably small the selective approach of not recommending large-scale land based projects is being followed widely. In some of the islands the land-based projects are suggested mainly to improve the infrastructure facilities. Also many of the recommendations considered as part of the perspective developmental plan fall within the ambit of improving the existing situation with appropriate policy inputs and corrective measures.

Tourism in the islands could be developed as a priority through grouping by considering the existing infrastructure facilities and distance between the islands. Four groups are identified viz., Kadamat, Amini, Chetlet, Kiltan and Bitra (Group I), Kavaratti, Agatti, Bangaram and Tinnakara (Group II), Androth, Kalpeni, Thilakam, Cheriyam (Group III) and Minicoy and Viringili (Group-IV). The available land area outside the CRZ that can be used for tourism in each island also has been identified and shown in the ICZMP map.

6.5.7 Environmental Education for People

The developmental projects proposed in the ICZM plan requires environment assessment processes for the possible impacts on the coastal resources. Some recommendations are made as part of the ICZMP to reduce the negative impact. The high density of dwelling units in certain islands very close to the coast has affected the environmental quality. The limited land availability and population pressure have put enormous pressure on the island resource. The present population of approximately 64,000 [11] is expected to double at the present day growth by the year 2025. Even if there is a reduction in the rate of growth, there are indications that the one lakh population mark shall be crossed by the year 2025 [12]. Hence there is an urgent need to educate the people to have a small family considering the limited resources they possess.

6.6 Conflicts and Perceptions of the Stakeholders in Islands

The major areas of conflicts concerning the islands are the over utilization of ground water resources and coastal erosion that is attributed to the developmental activities such as harbour development, mining of corals, etc. In addition to these activities there are other areas of conflicts such as disposal of sewage and solid materials, degradation of corals, depletion of bait fish, shore protection, etc. which are being worked out to decrease the impacts and also to enhance the resources. These are appropriately integrated during the preparation of implementation schemes for the islands.

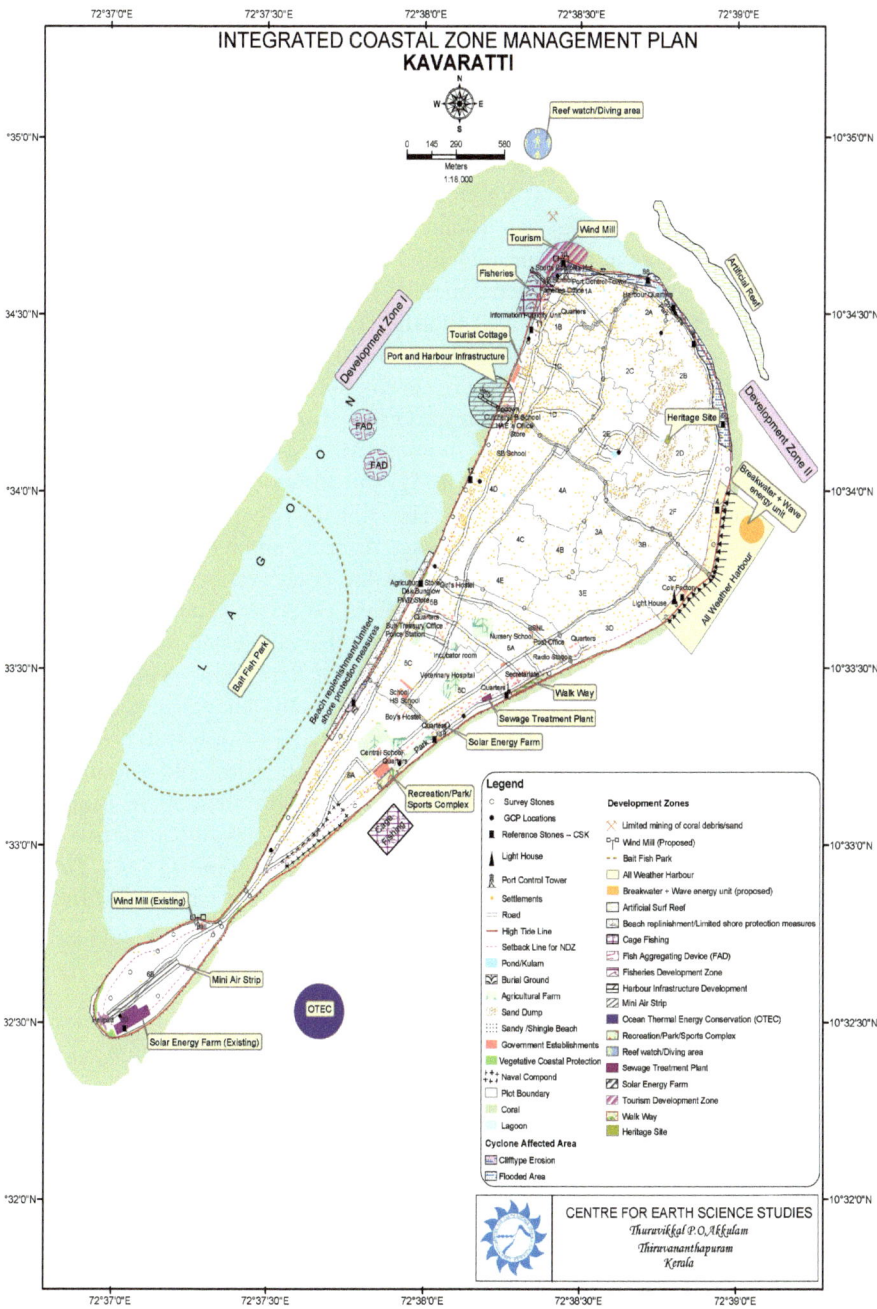

Fig. 6.1 ICZM plan for Kavaratti Island

6.7 ICZMP

ICZM is accepted as a blueprint for the sustainable island development. The success of ICZM depends on the efficiency in the implementation of the action plans. Effective implementation is an evolutionary process requiring a combination of management initiatives with optimum technological interventions. A good mix of administrative policies and island council initiatives and technological/scientific inputs and resource management is the important ingredient that would lead to effective execution. The preparation of plan is a challenging responsibility considering the quantum of work and logistic constraints for data collection and stakeholder interaction in the island. For the first time, ICZMP has been prepared for all the inhabited islands of Lakshadweep including the uninhabited island of Bangaram. A sample ICZM Plan of Kavaratti Island is given in Fig. 6.1. Majority of the islands have an elevation of 3–4 m above MSL. Patches of high sand dunes with elevations more than 5 m above MSL are observed in Kavaratti, Amini, Androth, Agatti and Minicoy islands. The Lakshadweep Administration has already implemented a project for safe sewage disposal and a desalination plant as recommended in the ICZMP. As per this plan the conservation, preservation and development zones are identified and action plans proposed. The fishery resource exploitation and catch enhancement schemes are also suggested through a failure analysis study. The Lakshadweep has a reef area of 816.1 km². The state of the coral reefs in all the inhabited islands was ranked in an abundance scale ranging from very good to good, satisfactory, unsatisfactory, and from bad to critical [13]. Based on the results it has been concluded that Agatti, Bitra and Kiltan islands have very good corals with a live coral abundance of more than 30 %. Bangaram, Chetlet, Kavaratti and Suheli islands fall under the good category with live corals accounting to more than 20 %. Amini has satisfactory live corals (15 %). The corals of Androth, Kalpeni and Minicoy are in the unsatisfactory category of live corals with a low abundance of 10–15 %.

6.8 Summary

All the conservation, preservation and development zones in the islands are identified and action plans proposed. The Lakshadweep administration can use the ICZMP as an effective tool to plan their developmental activities without affecting the ecosystem. Desalination plants of 1 lakh litre/day capacity have already been installed in the three major islands viz. Kavaratti (2005), Agatti (2008) and Minicoy (2010). The remaining islands will also have this facility shortly. The CRZ rules regulating the construction and development activity are in existence and this plan helps to redefine the No Development Zone (NDZ) under the notification. Adherence to the ICZM plan will help to sustain the sensitive coral ecosystem with minimum impact.

References

1. Clark JR (1995) Coastal zone management handbook. Lewis Publishers, New York
2. Report IREL (2002) Heavy mineral budgeting and management at Chavara. Centre for Earth Science Studies, Trivandrum
3. Integrated Coastal Zone Management Plan (ICZMP) for Lakshadweep Island (2006). Final report submitted to the Ministry of Environment and Forest (MoEF), Government of India, Centre for Earth Science Studies, Trivandrum
4. Environment Protection Act (1986) Government of India
5. Coastal Regulation Zone (2011) Ministry of Environment and Forest, Government of India
6. Island Protection Zone (2011) Ministry of Environment and Forest, Government of India
7. Coastal Regulation Zone (1991) Ministry of Environment and Forest, Government of India
8. Thomas KV, Prakash TN, Shahul Hameed TS, Raja D, Manoj B (2002) Demarcation of the HTL, LTL and no development zone along the islands of Lakshadweep (Phase-I). Report submitted to Department of Science and Technology, UT Lakshadweep, Centre for Earth Science Studies, Trivandrum
9. Prakash TN, Shahul Hameed TS (2004) Site inspection report (Cyclone-2004) in Lakshadweep Islands. Submitted to UT Lakshadweep, Centre for Earth Science Studies, Trivandrum
10. White AT (1988) Marine parks and reserves: management for coastal environments in southeast Asia. International Centre for Living Aquatic Resources Management Manila, Philippines
11. Anonymous (2011) Census of India, 2011. Office of the Registrar General and Census Commissioner, India
12. Master Plan of Lakshadweep Islands 2025 (2009). School of Planning and Architecture, New Delhi
13. UTL (2002) Bio-physical surveys of coral reef health during 1999–2002 in Lakshadweep Islands. Department of Science and Technology, UT Lakshadweep